Panorama of Deep Learning Based Recommender System: Development & Challenges

Panorama of Deep Learning Based Recommender System: Development & Challenges

Dr. Bam Bahadur Sinha
Dr. R. Dhanalakshmi

CWP
Central West Publishing

This edition has been published by Central West Publishing, Australia

For more information about the books published by Central West Publishing, please visit https://centralwestpublishing.com

Disclaimer

NATIONAL LIBRARY OF AUSTRALIA

A catalogue record for this book is available from the National Library of Australia

ISBN (print): 978-1-922617-54-5

Preface

"Learn from yesterday, live for today, hope for tomorrow. The important thing is not to stop questioning" — Albert Einstein

As a major medium for business and e-commerce transactions, the subject of recommender systems gained growing prominence in the 1990s. Web facilitated the collection of data and a user interface that could be used to non- intrusively recommend items. The widespread internet adoption has resulted in an explosion in the number of consumer choices. In modern e-commerce, entertainment and social media platforms, the users start to expect personalized content. By retaining a user's loyalty and satisfaction with personalized content, Recommender Systems (RS) provide a vital resolution to this problem. Recommenders are key tools in industry for improving user experience and promoting sales and services for many mobile applications and online websites. 80% of the movies seen on Netflix came from recommendations and 60% of the videos watched on YouTube came from the YouTube homepage recommendation. The subject matter of recommenders is extremely diverse as it allows you to make recommendations using different kinds of user preferential and user- requirement data. Collaborative filtering methods, content-based methods and knowledge-based methods are among the most known methods in recommender systems. These three approaches form the basic foundations of experimentation and exploration in RS. During the last few years, advanced methods for various data fields and contexts such as location, time and social information have been developed. Numerous furtherance for specific scenarios were suggested and methods were adjusted for different applications such as news recommendations, query log mining and computational advertising, respectively.

Conventional RS techniques often limit the expressive power needed for the modeling of complex interactions between users and items. They require extensive handmade attributes from subject matter experts. Deep learning identified considerable accomplishment in addressing problems with advanced cutting-edge outcomes that have been reported by the artificial intelligence community in fields like computer vision and linguistic processing for many years. The RS consists of a wide variety of data, such as behaviour pattern data (e.g., click times, duration of browsing, etc.), proceedings data (e.g., event), production informative data (e.g., content descriptions, item attributes, etc.), rating data (e.g. rating scores) and socio-economic data (e.g. user identity, domicile, etc.). Deep learning can automatically capture the dynamic structure of consumer expectations by encoding high dimension data in the learnt function representations. The ground-breaking technology of deep learning has received considerable interest in speech recognition, image interpretation and natural language processing. Unlike conventional recommendation structures, deep learning efficiently captures non-linear and non-trivial interactions with users and allows more complex abstractions to be coded as data representations in the upper layers. In recent years, there has been an unprecedented growth in research publications on methods for profound learners, which demonstrate the unavoidable generality of deep learning while proposing any recommender system. The structure of book demonstrates the impact of deep learning on recommender systems. The chapters of the book can be assembled into two categories:

1. Omnipresence of deep learning in specific domains of recommender system: These chapters address deep learning techniques in recommender systems, including recommendation with deep learning techniques in content based system (Chapter 3), recommendation with deep learning techniques in collaborative system (Chapter 4), recommendation with deep learning techniques in hybrid system (Chapter 5), recommendation in context- aware systems (Chapter 6), and combination of social network & Trust- aware recommender system with deep learning(Chapter 7).
2. Advancement and application of deep recommender system: Chapter 8 is primarily aimed at providing the readers with basic ideas and principles driving trends in recent years. Although all the recent innovations cannot be addressed in depth in a single book, the content in the closing chapter of the book is intended to perform the role of "ice breaking" in advanced topics. The chapter further discusses other application scenarios using recommendations technologies, such as news recommendation, and computational advertising.

Since this book is ostensibly written as a textbook, it is understood that a significant part of the audience will be industry experts and scholars. Thus, we have taken care to compose the book content in such a way that it is always valuable from an applicable and research point of view.

Acknowledgements

First and foremost, I would like to thank my co-author, Dr. R. Dhanalakshmi who encouraged me constantly throughout the period of writing with her invaluable feedback and input. I express my deep sense of gratitude from the core of my heart, for her valuable guidance, suggestions and supports without which this book would not be in its present form. She has been my inspiration and motivation for continuing to improve my knowledge and helping me move ahead in my career.

Most importantly, I acknowledge the unlimited love, encouragement, assistance, affection, support and blessings received from my mother, father, and brother, without which it would not be possible to write this book.

Bam Bahadur Sinha

Table of Contents

CHAPTER 1

Introduction to Recommender Framework and Deep Learning

"People influence people. Nothing influences people more than a recommendation from a trusted friend. A trusted referral is the holy grail of advertising" — Mark Zuckerberg

1.1 Framework of Recommender System

A framework that performs recommendation can be considered as a search ranking system, where the input parameter is a collection of users and context, and the output is a ranked list of products. When a query is fired, the recommender system searches for the relevant items in the system database and then classify items according to certain strategic objectives, such as click event or purchase event. Recommender systems (RSs) guesstimate customer preferences for products and proactively recommends certain products that the end users might find interesting [AT05]. The key concept is to examine the history and the preferred details of the user and construct a model to suggest the service or products of interest automatically, and then to obtain a customized list for the consumer. Not only can the framework of recommender system suggest products that have common consumer experience, but it can also suggest unknown products of interest to the consumer without knowing the preference of the consumer. These challenges may be solved by effectively defining the possible desires of consumers and choosing the appropriate products from a broad variety of consumer data. As websites for e-commerce continued to grow, it became desperately important to include recommendations focused on filtering the full spectrum of alternatives accessible. Consumers find it challenging to get the right alternative out of the wide variety of items (products and services) provided by this website. Recent years have witnessed a drastic rise in the importance of recommendation systems, as the following evidence indicates:

1. In high-ranking online platforms such as Netflix, Amazon, Facebook, Last.fm, YouTube, Pinterest, IMDb, Twitter, Spotify, TripAdvisor, Instagram, etc., feedback services play a sig-

nificant role. In fact, several media outlets are already creating and distributing recommender systems to support their consumers. Netflix, the on-demand digital service company, for example, awarded the team with a $1 million award, which also significantly increased their recommender system's output.

2. Specifically, there are seminars and workshops on the field: the Conference Series on Recommendation Systems (RecSys), entrenched in 2007 by the Association of Computing Machinery (ACM). This conference is the first international occurrence to suggest study of RSs and innovations related to RSs technologies. Throughout the field of computer management and adaptive structures, workshops devoted to RSs are most frequently taken up in more conventional conferences. Many prominent workshops within this area include: SIGMOD - Special Interested Group on Data Management of ACM, SIGIR - Special Interest Group on Information Retrieval, WWW - World Wide Web, and UMAP - User Modeling, Adaptation and Customization.
3. Several graduate courses are now entirely devoted to RSs, tutorials on RSs are very common at computer science conferences at higher education institutions around the world, and many books on RSs has been already published [Agg+16][RRS11][TRD20][10]. Recently, a broad new set of articles was published that aims towards development of recommender framework for software applications [Rob+14][NBF19].
4. In academic journals there were several special issues covering research and development in the field of RSs. Journals of specific problems related to RSs include: ACM Transactions on Computer Human Interaction, ACM Transactions on Intelligent Systems and Technology, ACM Transactions on Information Systems, ACM Transactions on Interactive Intelligent Systems, IEEE Intelligent Systems, International Journal of Computer Science and Applications, etc.

Indeed, many explanations exist for the usage of this platform by several service providers:

1. Boost up the sale of products: It could be the main feature for commercial RSs, i.e. to provide an extra collection of products available without any form of suggestion compared

with those normally available. This goal is accomplished as the products suggested are generally relevant to the needs and preferences of the customer. The consumer will probably recognize it after several recommendations have been tried. In general, from the point of view of the service provider, the key aim of adding RSs is to enhance the conversion rate, i.e., in contrast with the number of visitors who merely search the content, the number of users that follow the suggestion and consume any product.

2. Boost up customer satisfaction: The customer will consider the feedback informative, relevant and will also appreciate utilizing the program with a well-designed human-computer interface. A properly built RS will enhance the user's application or web experience as well. The balance between efficiency, precision and a functional graphical user interface (GUI) improves the subjective assessment of the system. It would therefore improve program usage and the probability of adoption of the recommendations.
3. Boost up sale of diverse products: One of RSs 'key function is to enable the consumer to pick products without a clear request, which may be challenging to locate. In a tourism sector, for example , the service provider wants to encourage not just the most famous but all of the places of interest in a tourism city. That may be challenging without a RS, as it is impractical for the service provider to avoid advertising somewhere not expected to match the preference of an individual customer. Thus, RSs performs the job of offering unpopular places to correct users or advertises them.
4. Boost up user loyalty: A customer will stay faithful to a website that acknowledges and respects an old client as a trusted customer while visiting. This is a typical function of an RS because many RSs utilize the data obtained by users through prior experiences such as the user's reviews, purchase data, rating scores, etc. The more the consumer engages with the interface, the more the recommendation quality are refined: the framework is able to reflect the needs of the user and the quality of the suggested performance is improved to adjust and suit the expectations of the user.
5. Increased awareness of what the consumer needs: The definition of user expectations, which can be obtained directly or projected by the application, is another valuable feature

of any RS and can be leveraged for many other applications. Such expertise can then be utilized by the service provider for several other purposes such as optimizing resource control and development. In the travel domain, for example, destination managers can decide whether to advertise specific areas to new customer sectors, or advertise certain types of promotional information that are derived from an analysis of the data collected by the RS (user transactions).

The recommendation framework is described as a tool for decision-making in complex knowledge environments. Some of the formal definition for recommender system used by several researchers are as follows:

[Bur02], 2002:

Recommender system is a framework that provides personalized sug- gestions as a result or that guides the consumer directly to useful or fascinating items from a wide variety of potential alternatives. - 2002

[Liu18], 2018:

The recommendation framework can be classified into three principle components namely: user-based, product-based and recommendation algorithm. The user-based model is usually used to collect, view and store data of users, the development design of product-based model applies to the product attributes and the recommendation algorithms primarily determines the tastes & desires of the customer by leveraging the customer's past preference. - 2018

[DCK20], 2020:

The framework of recommendation addresses the knowledge complexity problem and help users select products from available alternatives. RSs gather information related to users' interests and generate an appropriate product suggestion to please users. - 2019

1.2 Input Data and Knowledge Sources of RSs

Dependent on the specific situation, a kind of recommender is selected. This scenario is primarily characterized by the various information outlets, which are used for the advising method as input variables. Following are the different types of knowledge sources used by recommender systems:

1. Customer Information: The framework needs comprehensive user knowledge to be considered as user model in order to establish acceptable recommendations. These can provide personal or preferred details for the customer.
2. Catalog Information: This collection can be very complicated and redundant based on the context, as this knowledge is displayed as characteristics, descriptions, branding or tags. This information represents the profound awareness of the suggested product.
3. Serviceability Information: The system aims at fitting the desires of the user with specific set of items. It stores the method or the type of algorithm used by the system for performing the pattern matching.

It is really necessary to be able to learn what consumers search on the site in order to build the consumer portfolio. For the estimation of the rating, different websites consider different sorts of input data from the consumer. The input details [SD19b] can be categorized into five groups as discussed below:

1. Social and Economic Data: This provides comprehensive details on a population's characteristics such as customer id, gender, education, profession, contact numbers, date of birth, residence, hobbies, job experience and income.
2. Observable Data: This type of information is used to define growing patterns of communication between products and to enforce patterns. It includes click time periods, browsing durations, web content user operation, e.g. retrieving, scrolling and others.
3. Ranking/Rating Score Data: The perceived quality of any product basically reflects a rate. This contains multiple reviews such as discrete and continuous scores; underlying observations such as terrible and great.
4. Proceedings Informative Data: It defines the occurrence of any event with certain time stamp values associated with it. It provides the sales information such as date of purchase, sales number, quality, quantity and discount.

5. Production Data: This type of data is maintained on an ongoing basis for business process control. Market managers utilize details including user specifications for online movies and product features to take reasonable actions that can match economic success of the company. In order to maintain its usefulness for the enterprise, production details must be properly documented.

1.3 Stages of Recommendation Process

Recommendation process involves three basic stages as depicted by Figure 1.1.

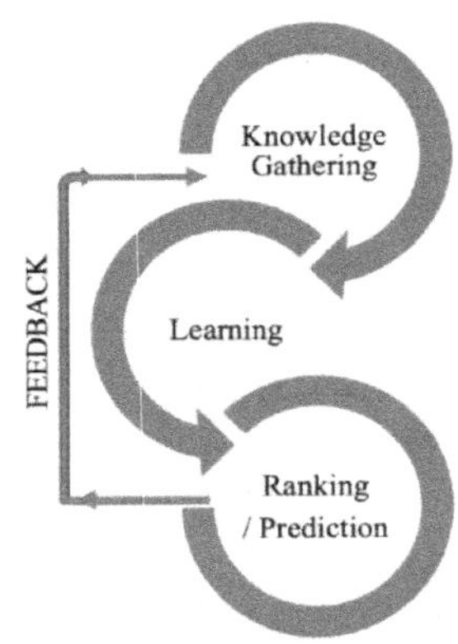

Figure 1.1 Basic Stages of Recommendation Process

Knowledge Gathering Stage

It gathers the necessary user details including the feature of the user, the actions or the nature of the services that the user accesses to create a user profile or model for predictive activities. The recommender agent cannot work precisely unless the user profile/model has been well built. The recommender program wants to learn as much as possible from the customer in order to make a sound suggestion right from the outset. Recommendation mechanisms depend on various forms of data, such as the explicit feedback [MM20] of highest standard, which involves users' explicit advice on an entity value or implicit feedback [Wan+20b] that analyses consumer behaviour indirectly for deciding their preference. Although explicit feedback calls for increased effort on the part of the user, more reliable data still relates as the preference is not taken from actions. Implicit feedback turns down the burden on the user by analyzing their behaviour indirectly. This approach of feedback generally tends to form an inaccurate recommendation. Few researchers also made use of both explicit and implicit feedback. Such feedback are referred as hybrid feedback [CL20]. A user profile is a list of personal details linked to a single individual in the e-learning network. It involves thinking abilities, analytical capacity, behavioural patterns, desires and system participation. Typically, the user profile will collect the details required to construct a user model. A user profile then represents a basic user interface. The effectiveness of every proposed recommendation model primarily relies on the willingness to serve the actual needs of the customer. In any statistical technique, reliable models are necessary to achieve appropriate and correct recommendations.

Learning Stage

In this stage of recommendation process, a learning algorithm is applied for the filtering and exploitation of user attributes which were collected during the stage of knowledge gathering.

Ranking/Prediction Stage

It recommends or forecasts the sort of products in which the consumer might be interested. It may be accomplished either explicitly based on the dataset gathered during the memory or model-related knowledge gathering stage or from the observable behaviours of the users present in the system.

1.4 Taxonomy of Recommendation System

In this section, we have discussed the potential of different recommendation techniques. Each technique has some pros and cons associated with them. Figure 1.2 illustrates the anatomy of different traditional filtering techniques.

Content-Based Filtering (CBF)

It is a domain-driven algorithm and emphasizes the study of the characteristics of products to generate forecasts. The suggestion is rendered using the consumer profiles of content-based filtering strategies utilizing functions derived from the contents of the products analyzed in the past [Bob+13]. After evaluating an adequate number of products that have already been preferred by any individual, a profile of user interests is created. The suggestion method will then scan the collection and pick appropriate products on basis of the generated profile [AN20].

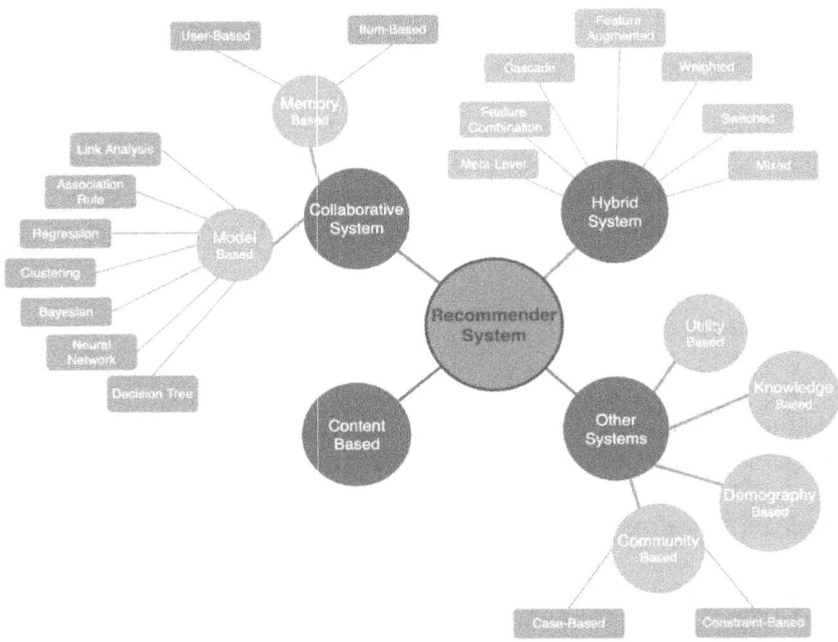

Figure 1.2 Taxonomy of Recommendation System

Figure 1.3 depicts the working of CBF. In contrast to other recommendation framework methods few strengths and drawbacks of the content-based method are as follows:

Pros of CBF:

1. By analyzing product characteristics and user profile, the recommendation list can be generated.
2. It can avoid cold start [Nat+20] item problem by recommending even the newer items to the user based on their profile.

3. This system tells us about the characteristics of the products which are the foundation of the recommendation. It reflects transparent functioning of generated recommendation list.

Cons of CBF:

1. Lacks diversity and novelty in the recommended set of products.
2. Requirement of domain knowledge prior to building of user profiles.
3. If the items lack sufficient attribute information [GLD18], it would be a difficult task for CBF to generate recommendations.
4. Problem of Filter-Bubble [Ngu+14]. It states that the user will get recommendation of only the similar items in which he/she has shown interest in the past.

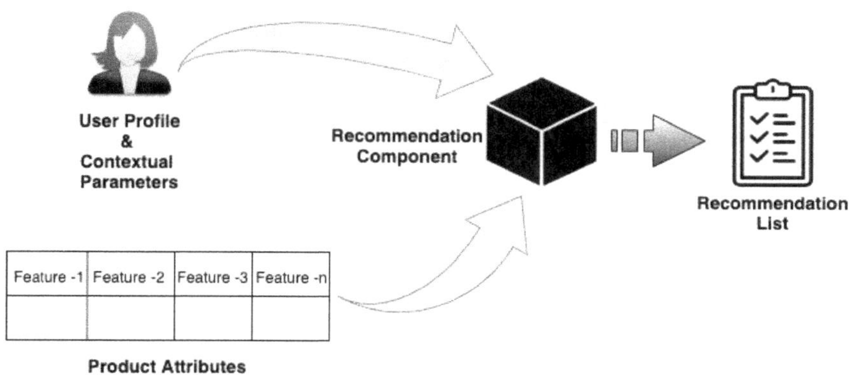

Figure 1.3 Content-Based System

Collaborative Filtering (CF)

Collaborative filtering suggests products that are focused on the preferences of other users or recommend products similar to those that the targeted consumer has already rated. Two consumers have a distinction in their preferences depending on the similarities in their past rating data. This is the most widely used recommendation technique and it is also referred as "people-people correlation"[SKR01]. Figure 1.4 depicts the working of collaborative filtering. The collaborative filtering technique can be subdivided into two categories namely: Memory-Based and Model-Based.

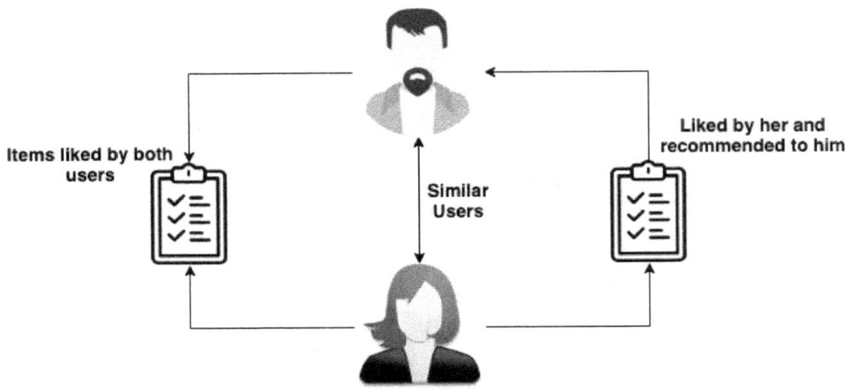

Figure 1.4 Collaborative Filtering

1. **Memory-Based:** The products previously assessed by the consumer play a significant role in looking for an individual who shares his admiration. So, when a user's neighbour has been identified, different algorithms may be used for integrating neighbor's interests to generate recommendations. Because of the usefulness of these approaches in real life implementations, they were quite popular. It can be successfully achieved by two approaches namely: User-Based and Item-Based. User-oriented strategies measure correlation between users by contrasting their ranking to the same product and then determine an active user's expected ranking for any product as a weighted average of the active user scores where weights are identical to the target item. Item-Based strategy generates a corresponding item- item similarity model by extracting all items from the user-item matrix that are evaluated by an active consumer, calculates how close the items are to the targeted item, then selects the 'k' most related items. Final prediction is made by taking a weighted average of rating score given by active users to the 'k' selected items. Several similarity techniques such as Pearson Correlation Coefficient (PCC), Cosine Similarity, Adjusted Cosine Similarity, Jaccard Similarity, etc. [SD19a] can be used by any recommendation model.
2. Model-Based: This approach utilizes previous evaluations to learn a model and enhance the efficiency of CF. Several deep

mining, machine learning or deep learning techniques can be used for learning and creating the model. Such strategies can easily suggest a variety of products as they utilize a pre-computed model which has shown outcomes that are close to recommendation techniques based on neighbourhood approach. Some of the well-known learning algorithms used by the model based approaches are as follows:

- Association Rules: It pulls out the rules that forecasts the presence of any particular item based on the presence of other items in the transaction. The representation of preferences can be very compact, which can improve both storage efficiency and performance. Furthermore, the association rule is also known for its efficiency in uncovering trends and guiding custom marketing decisions [Muh+20].
- Bayesian: Bayesian classifiers may be effective in environments where user expectations shift gradually with respect to the time needed for model creation, however they are not ideal for situations where user favored models need to be updated rapidly or regularly. Even in models-based RSs, it is popular because it is also used to generate a framework for CBF recommendations. Bayesian classifiers are based on Bayes theorem and conditional probability which aims to solve the classification problem [IS20].
- Clustering: Clustering algorithms aim to fragment a collection of data into a number of sub-clusters to identify important classes among them [Yan+20]. Once the clusters are formed, the opinions of other users can be averaged in a cluster and used to recommend for each user. An efficient form of clustering results in better quality clusters in which there is a strong intra-cluster similarity although the inter-cluster similarity is weak. In certain clustering strategies, participants in multiple clusters will partially take part, and decisions are instead focused on the total of participating clusters that are determined by their degree of involvement [LSY03]. K-Means, Fuzzy C-mean, and Self Organizing Map are some of the well-known clustering approaches being used to reduce the candidate set in CF based systems.
- Decision Tree: It is focused on the tree graph approach and is centered on an overview of a collection of training examples with known class labels. These are then used to categorize the previously encountered instances

[Ger+10]. They will render very reliable predictions if training of decision trees is performed on high quality data. In managing products with a mix of real-valued, category attributes, and things that have certain unique characteristics, decision trees are often versatile.

- Link Analysis: It is the method to create linked entity networks such that patterns and developments are examined[Par+12]. It has shown tremendous potential to boost web search efficiency. The web page is a single node of the network graph in most link analysis algorithms. Link analysis comprise of two sub-parts namely: Hyperlink-Induced Topic Search (HITS) and PageRank algorithm.
- Neural Network (NN): It is a network composed of several linked neurons (nodes) organized systematically in layers. The interaction between neurons has weight based on how strongly one neuron affects another [Par+12]. In some specific issues, the usage of neural networks gives several advantages. Eg: Since it comprises several neurons and the weight of each connection, a neural network approach is very resilient in terms of noisy and faulty datasets. NN can approximate nonlinear functions and can prove to be efficient in reporting complicated relations in datasets and it can also operate when a network component fails. The only downside of NN is that it's impossible to determine the optimal network topology for a given problem, which would serve as the lower limit for the classification error until the topology is settled on.
- Regression: It is used where a causal connection assumes that two or more variables are consistently related. It is an efficient, diverse method for evaluating the relationship of dependent variables with one or more independent variables [WL14]. Regression applications involve curve fitting, estimation, and rigorous monitoring of relationships among the variables. The curve can be helpful for defining a pattern, be it linear or parabolic, with other types, inside the dataset.

In contrast to other recommendation framework methods few strengths and drawbacks of the CF:

Pros of CF:

1. It analyses change in user behaviour.
2. It has the potential to generate serendipitous and diverse set of items
3. For large user spaces, the collaborative system shows significantly high performance.
4. It addressed the problem of "filter-bubble".
5. No domain knowledge is needed in the initial recommendation phase

Cons of CF:

1. Suffers from the problem of cold start.
2. In case of high dimensional dataset, the complexity of CF system increases.
3. Sparse user-item matrix leads to inability in locating potential neighbors and thus makes weak recommendations.

Hybrid System

Hybrid filtration incorporates different approaches in order to configure the system further, to avoid the limitations and complications present in pure RSs. The theory behind hybrid approaches is that a mixture of algorithms would be more efficient and accurate than one single algorithm, as the drawbacks of one algorithm can be solved by another. A hybrid model can remove the shortcomings of any particular strategy with the usage of multiple suggestion techniques. The integration of methods may be carried out in many ways: independent algorithm deployment and comparison of the performance, utilizing a certain degree of content-based in CF, utilizing collaborative filtering in CBF applications, providing a single recommendation framework that incorporates all methods, etc. Following are some of the well-known hybridization approaches:

1. *Weighted:* The ranking of different recommendation schemes is compiled into a single recommendation. The findings are measured according to their effect on a particular item when integrating them. Eg: P- tango[Mir+99] consist of a hybridized model that combines CF and CBF approach of recommendation. Initially, both systems are assigned with

identical weights, and then the weights are adjusted in further steps by analyzing the predictions made by the hybridized model.

2. *Cascade:* In constructing an order of choice between various items, the cascade hybridization approach implements an iterative mechanism of refining. The recommendations of one methodology are further improved by another. The first technique provides a provisional set of suggestions which is improved by the next methodology. Regardless of the coarse to finer aspect of the application, the hybridization process is very effective and noise resistant. E.g.: EntreeC [Bur02].
3. *Feature Augmentation:* It is a form of sequential recommendation where output of one Rs is used as an input for another RS. It also has the capability of appending new features and functionalities in the second attached RS. E.g.: Libra System [MR00]
4. *Feature Combination:* Different feature aspects of the information base are clubbed into a single domain. For example, demographic information and predefined utility factors may cover these features. Pipper [BHC+98] is a system that made use of collaborative filter rating data in CBF system as feature for making recommendation of certain movies.
5. *Switched:* The recommending program changes between different forms of recommending systems according to the situation and chooses the one that is most suitable for the specific issue. The positive side of this approach is that the model is responsive to the constituent recommenders' strengths and weaknesses. One of the major demerits of switched systems is that it increases the model complexity and also increases the burden of RSs for determining the increased number of parameters present in the system. E.g.: DailyLearner [BP99] employs CBF for addressing cold start problem and uses CF system otherwise.
6. *Meta-Level:* In case of meta-level systems, before actual deployment of the model, the primary focus is on understanding and analyzing the meta structure of the snag. E.g.: LaboUr [SKK01] utilizes instant learning methods to build user profiles which is further compared collaboratively. The meta-level-based models possess the potential to deal with sparsity issue being faced by CF systems. It uses the entire

pattern that the first RS model has learned as a reference for the second RS model.

7. *Mixed:* It performs parallel implementation of various forms of recommendation programs and reports the test results concurrently. The customer is provided with all relevant recommendations depending on different methodologies. E.g.: PTV [SC00], Profinder [Ahm98], and, PickAFlick [BHY97].

Other Personalization Systems

1. *Demography-Based:* The basis for this form of recommendation method is to identify the various demographic attributes such as age, class, income, cultural history or other personal attributes [LKH14]. The generated model is then paired with a manually produced list of consumer expectations with all associated personal attributes. Demographic structures may be used in a number of ways, but in particular it helps in addressing the cold start user problem (new users). As the preferences of users are not identified, only demographic details from existing users may be used to render recommendations.
2. *Utility-Based:* Such programs seek to measure the value of every item for the consumer. In the next stage, they seek to increase this usefulness factor in order to satisfy increasing user's specific needs. A long-term customer assessment is not needed, because the program learns the consumer over time. The key challenge is to identify for each user this feature, which is always achieved through numerous customer satisfaction techniques [Den15].
3. *Knowledge-Based:* The fundamental knowledge regarding the connection between a user's desires and a particular product resides in a knowledge- based recommendation framework. The correlation between user requirements and a certain product characterizing all forms of recommendations may be appended in the user profile. In general, any form of expert structures should be viewed as a model dependent on knowledge, since the meaning of the program is generated by acquiring the user's information [Car+12].
4. *Community-Based:* This kind of advisory program forms a community possessing shared interest. It requires involve-

ment of an individual in a community and proposes item after the community has provided the decision [FB11].

1.5 Evaluation Metrics for RSs

Specific forms of accuracy or coverage may be used to determine the reliability of a preferred recommendation algorithm. Depending on the filtering technique, the method for measuring the recommendation quality varies. Accuracy calculation approaches for recommended filtration systems are split down into two approaches namely: statistical and decision support approach. The appropriateness of each metric depends on the characteristics of the dataset and the form of activities that the recommender program performs.

Statistical Approach: It assesses the performance of a filtering methodology by specifically contrasting the forecasted rating with the actual rating given by the user. Mean Absolute Error (MAE) and Root Mean Square Error (RMSE) are the two most commonly used technique for measuring the deviation between actual and predicted rating. The mathematical representation for MAE and RMSE is given by Eq. 1.1 and Eq. 1.2 respectively.

$$MAE = \frac{1}{n}\sum_{x,i} |F_{x,i} - A_{x,i}| \tag{1.1}$$

$$RMSE = \sqrt{\frac{1}{n}\sum_{x,i} (F_{x,i} - A_{x,i})^2} \tag{1.2}$$

In the above equations, 'n' denotes the total number of predictions made. $F_{x,i}$ denotes the predicted rating and $A_{x,i}$ denotes the actual rating given by user 'x' to item 'i'. RMSE highlights large absolute errors. Lower MAE and RMSE value indicate that the model is performing better.

Coverage reflects the percentage of recommended products from the total possible alternatives. It is generally classified into four categories namely: Catalog Coverage, Item Coverage, User Coverage and Coverage of User Interactions. Item coverage calculates the percentage of items recommended from total set of available items. User coverage denotes the percentage of users for which recommendation has been generated from total possible set of potential users present in the system. Catalog coverage measures the percentage of recommended user-item pair over total possible user-item pair in the system. Coverage of user interaction denotes the percentage of predicted ratings over total number of ratings.

Decision Support Approach: The metrics see the estimation process as a quantitative activity choosing between useful items and weak items. Such criteria allow consumers to choose good quality products from a wide range of accessible items. Some of the commonly used decision support metrics are: Receiver Operating Characteristic (ROC), Precision, Recall, F-measure, Weighted error, and Reversal Rate. The mathematical representation of Precision, Recall and F-measure can be given by Eq. 1.3, Eq. 1.4 and Eq. 1.5 respectively.

$$\text{Precision(P)} = \frac{\text{Correct Recommendation}}{\text{Total recommendation}} \tag{1.3}$$

$$\text{Recall(R)} = \frac{\text{Correct Recommendation}}{\text{Total useful recommendation}} \tag{1.4}$$

$$\text{F} - \text{measure} = \frac{2 * P * R}{P + R} \tag{1.5}$$

1.6 Deep Learning for RSs

Deep learning approaches are designed to acquire hierarchies with characteristics from higher hierarchies formed by the compilation of characteristics at a lower level. Automatically learning features at several abstract levels allow the machine to learn complicated functions, which explicitly map the input to the output, without entirely depending on human-crafted functions. In general, deep learning is a branch of machine learning. The fundamental meaning of deep learning is that it studies complex representations, i.e., several layers of data interpretation and abstraction. For functional justification, we find every neural architecture as a "deep learning," such that it optimizes a differentiable objective function with a stochastic gradient descent (SGD) version. In supervised and unsupervised learning functions, neural architecture has proved tremendous progress and success.

Exigency of Deep Learning for RSs

Before you immerse yourself in the specifics of recent developments, it is important to consider why recommender frameworks are being advanced through deep learning techniques. In past few years, it is apparent that several deep recommendation frameworks were introduced. At this stage, the necessity for too many deep learning based recommender frameworks or even even the usage of neural networks for the recommendation domain can possibly be debated. In the same manner, it should offer a straightforward explanation of whether and under which situation suggested design will profit the most. In brief, this concern is very important for the function, frameworks and scenarios of the recommender system. The strengths of the prediction models focused on deep learning that readers should take into consideration when attempting to use them for recommendation are as follows:

1. *Flexible Behaviour:* Deep learning techniques are very versatile, in particular with the arrival of numerous popular deep learning structures like Caffe[CÇÖ17], DeepLearning4j[Lan+19], Keras[GP17], MXnet[Che+15], Microsoft Cognitive Toolkit[Chu+17], Neon[Bah+15], PyTorch[Bah+15], Tensorflow[Chu+17], Theano[Chu+17], etc. Most of these methods are modularly built and fully endorsed by the research professionals. The successful modularisation increases the efficiency of technological innovation. For example, to devise efficient hybrid modules or substitute one module with another, it is simple to combine various neural structures. We may thus effectively construct hybrid and composite advisory models to capture various characteristics and factors simultaneously.
2. *Modelling Non-Linear Data:* In comparison to linear models, the recommendation model based on Deep Neural Network (DNN) can make use of non-linear functions such as tanh, relu, etc. to model non-linear data. This function enables the complex user experience trends to be observed in a dynamic way. Neural networks can estimate all continuous functions

with arbitrary precision by varying choices and combinations of activation [Hor91].

3. *Feature Learning:* DNN possess great power in learning the basic factors and functional representation displayed by the input data. In addition, in real-world implementations there is a vast volume of detailed knowledge regarding items and users. Having this knowledge is a means to improve our interpretation of items and users, and contributes to a stronger decision. As such, it is common to use DNN in recommendation models for feature learning. Two fold benefits of using DNN to assist feature learning are: a) DNN enables automated, unsupervised or supervised feature-level learning from raw data. b) It allows the inclusion of heterogeneous contents such as audio, video, etc.
4. *Sequence Modelling:* DNN have demonstrated impressive results in a variety of computational functions, such as machine translation, interpretation of natural languages, voice recognition, chatbots, etc. Recurrent Neural Network (RNN) and Convolutional Neural Network (CNN) are two widely acceptable deep learning techniques for extracting sequential information from the raw data.

The diverse set of architectural designs based on deep learning are as follows:

1. *Autoencoder (AE)* works in an unsupervised fashion. It tries to reconstruct the input data at the output layer. It has several variants such as: Sparse autoencoder (SAE), Denoising autoencoder (DAE), Contractive autoencoder (CAE), and Variational autoencoder (VAE).
2. *Adversarial Networks (AN)* is a neural generative network made up of a discriminator and a generator. Both neural networks are concurrently trained.
3. *Attentional Models (AM)* neural architecture are focused on soft content that addresses a series (or image). It is one of the emerging trends in deep neural network based recommendation research.
4. *Convolutional Neural Network (CNN or ConvNet)* comprise of convolutional layers that captures the local and global characteristics, thus enhancing the efficiency and accuracy of the

network. It yields very high performance in case of grid-like topologies.

5. *Deep Reinforcement Learning (DRL)*: The synthesis of deep neural networks and augmented learning formulates DRL, which improve human success in many fields such as games and auto driving.
6. *Multilayer Perceptron (MLP)* a.k.a. universal approximators comprise of stacked layers that models the association between input and output. They are often applied to the supervised problems. Backpropagation adjusts the weight and bias value of the network.
7. *Neural Autoregressive Distribution Estimation (NADE)* leverages the product rule and weight sharing strategy inspired from RBM. It is an unsupervised approach based on feed forward neural network and autoregressive model.
8. *Recurrent Neural Network (RNN)* aims at analyzing and modelling the sequential data. In order to keep track of former computation, RNN comprise of loops and memories. Its variants include Long Short Term Memory (LSTM) Classic, Peephole Connections, Gated Recurrent Unit (GRU), Multiplicative LSTM, and LSTMs With Attention.
9. *Restricted Boltzmann Machine (RBM)* restricts the interaction between the hidden layers or visible layers. It has two layer architecture and can easily be stacked as deep network.

Drawbacks of using Deep Learning for RSs

In this section, we will address some specific objections against the usage of deep learning for RSs.

1. *Inexplicable:* Notwithstanding its popularity, deep learning is renowned for being black boxes, so it seems to be genuinely a challenging task to provide explanatory predictions. A prominent claim against DNN is the non-interpretable, limited description of hidden weights & activations. However, the introduction of neural focused models [TLH18] made this problem simpler and prepared the way for DNN models with an improved interpretability. Although processing individual neurons still remains a problem (not just in RSs) for neural networks, existing state-of-the-art networks now

have some interpretability and are thus possible to render recommendation that can be explained.

2. Hunger for Data: A second potential drawback is the awareness that deep learning is data-hungry, because it requires adequate data to completely endorse its rich parameters. Nonetheless, it is fairly straightforward for labeled data to collect significant quantities of data in the form of proposing systems work in contrast with other domains (for example, language or vision). Million/Billion datasets are not only standard in the industry sector but are often made accessible as research datasets.
3. Substantial Tuning of Hyperparameter: We notice that the tuning of hyperparameters (eg: regularization factor, learning rates, etc.) is an overarching challenge of deep learning. This problem persisted with even the machine learning techniques.

1.7 Deep Learning in Specific Domains of RSs

In this section, we discuss the omnipresence of deep learning in several domains of recommender system. The RSs based on deep learning techniques can be broadly categorised into two groups namely: a) Neural Network framework, and b) Combination framework. The combination framework can be further divided into two subcategories: Combination of traditional RSs with deep learning techniques, and second category constructs model that solely uses the deep learning technique. Following are the diverse set of domains in which deep learning based recommender system are used:

1. Recommendation with deep learning techniques in content-based system: The primary of this approach is to efficiently capture the non-linear interaction between users and items present in the system. In addition, it extracts complex relationships within the data, such as historical, textual and visual knowledge, from ample open information sources.
2. Recommendation with deep learning techniques in collaborative system: In order to the recommendation model, it makes use of user information (explicit or implicit feedback) to train the deep learning model. The fundamental principle is to use a consumer rating matrix as an input, use the deep analysis model to know the feature representations of users

and items and to create an optimal system by using the loss function to refine the parameters of the deep learning approach.

3. Recommendation with deep learning techniques in hybrid system: It constructs a unified framework for making recommendation of certain items to the user. It combines the CBF and CF methods to form a hybrid model. The feature learning is done through deep learning approach.
4. Recommendation in context-aware systems with similarity technique based on feature weight learning: For several specific recommendations cases it can easily embed context information into recommender structures and using a deep learning approach it can acquire feature representations of contextual information. On this basis, the context-aware RSs can alleviate the problem of data sparsity and can integrate in all kind of rough situations.
5. Combination of social network & deep learning-based recommender system: It analyses the social relationship influence on the users. It models the socio-temporal pattern among the users and items to improve the performance of deep recommender model. But it has several weakness and limitations associated with it. In chapter 7, we tried to address some of the challenges being faced by the socio-temporal based deep learning RSs.

1.8 Summary

This book discusses the key aspects of the most important deep learning-based recommender systems. It provides a benchmark to several researchers and scientists to develop deep recommendation models for specific scenarios. We have explained the strengths and weakness of several domain specific deep learning-based recommendation systems. It is hoped that this book will provide the readers with a comprehensive information about notable advancements in this direction.

CHAPTER 2

Deep Learning Based Recommendation Models

"A little learning is a dangerous thing; Drink deep, or taste not the Pierian spring" — Alexander Pope

The framework of recommender system comprises of heterogeneous data such as unstructured contents (e.g., item descriptions), user-item data (e.g. purchase information), and item characteristics (e.g. movie genres). Deep learning technology has evolved rapidly in recent years, and has grown into an evolving and competitive area for knowledge discovery and machine learning. It can identify the dynamic structure of consumer expectations automatically via encoding high dimension data in trained representation of features. Recent developments in support of deep learning (DL) systems have been earned by addressing challenges to traditional models and attaining high quality recommendations. DL techniques achieved widespread success in the areas of image processing, object detection, natural language processing (NLP), and language comprehension. Deep learning has radically revolutionized the structures of recommendations and provided the potential to enhance the recommender's efficiency. It is capable of efficiently identifying complex and non-trivial user-item interactions and codifying more specific abstractions as high-level data representations.

2.1 Classification of DL-Based Recommender System

In order to provide an overhead view of DL-based recommender systems, we classified the models into two categories as follows:

1. *Neural Building Approach for Recommendation:* Autoencoder (AE), Attention Model (AM), Adversarial Networks (AN), Convolutional Neural Network (CNN), Deep Reinforcement Learning (DRL), Multilayer Perceptron (MLP), Neural Autoregressive Distribution Estimation (NADE), Restricted Boltzmann Machine (RBM), and Recurrent Neural Network (RNN) are the techniques that work on the principle of neural building blocks. These techniques help in determining the recommender system applicability. A detailed descrip-

tion of the aforementioned techniques is given in the upcoming section.

2. *Deep Hybrid Model Approach for Recommendation:* The versatility of Deep Neural Network (DNN) enables many neural blocks to balance each other and establish a strengthened hybrid model. There are several potential variations of these DL strategies, but not all of them have been explored. Some of the deep hybrid models are illustrated via Figure 2.1.

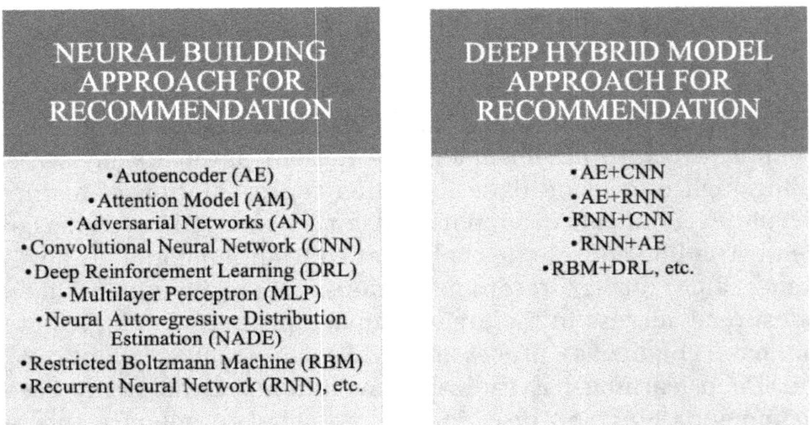

Figure 2.1 Classification of DL-Based Recommender System

2.2 Key Strengths of DL-Based Recommender System

The following are the four key strengths of DL-Based Recommender system as compared to tradition CF and CBF based recommender systems:

1. With the help of non-linear activation functions such as Sigmoid/Logistic, TanH/Hyperbolic Tangent, ReLU (Rectified Linear Unit), Leaky ReLU, Parametric ReLU, Softmax, Swish, etc., modelling of non-linear interaction in the selected data can be done using DL. Using this property, the complex and specific user-item interface patterns can be identified. Linear models such as matrix factorization approach limits the expressiveness of the recommender model and is very simplistic in nature. Neural networks have been well developed to approximate any continuous function by changing the acti-

vation preferences and combinations in an unpredictable manner. This property allows dynamic trends to be handled and represents the user's choice correctly.

2. DL can learn the valuable representations from data input effectively. Typically, there is a great deal of real-life implementations of detailed knowledge regarding items and users. Using this knowledge offers an incentive to enhance our interpretation of items and users, such that we suggest correct items to the consumers. As such, the usage of deep neural networks in representational learning in recommendation models is a reasonable alternative. Two-fold benefits of using DNN for representational learning are as follows: a) It diminishes the efforts required to design the features. b) Recommendation model allows the incorporation of heterogeneous information such as images, text, audio and video content.
3. DL is well suited for sequential modeling. RNNs and CNNs seem to be a powerful tool for handling tasks such as speech recognition, machine translation, etc. They are widely used and flexible for sequential data mining structure. Sequence signals modeling is an important topic for extracting the time fluctuations of the users' behaviour patterns and item advancements. E.g. Session based- RS, market basket analysis, etc.
4. DL possess high pliability capabilities. Following are the different DL frameworks that has community/professional support: Caffe[CÇÖ17], DeepLearning4j[Lan+19], Keras[GP17], MXnet[Che+15], Microsoft Cog- nitive Toolkit[Chu+17], Neon[Bah+15], PyTorch[Bah+15], Tensorflow[Chu+17], Theano[Chu+17], etc. To build efficient hybrid models, it is easy to merge different neural structures or to substitute one element with another. We could therefore easily create composite and hybrid recommendation models to collect different properties and factors simultaneously.

2.3 Recommendation Models Based on Deep Learning

In this section, we have discussed several deep learning techniques that has been used by different researchers to perform recommendation of products/items to the end user/consumer.

Multi-Layer Perceptron (MLP)

MLP [Hei+19] is a composite layer of complex nonlinear adjustments that learn hierarchical representation of features. It is a feed-forward network that comprise of multiple hidden layers between input and output layers of the neural network. The network will approx. any observable function, to any required precision. It is a succinct yet efficient network. It also acts as the foundation of numerous innovative methods and is commonly applicable in many other domains. In current recommendations program strategies, MLP can incorporate non-linear transformation and translate it into neural adjunction.

1. A recommendation can be defined as a two-way relationship between desires of consumers and functionality of items. Neural Collaborative Filtering (NCF) [He+17] and Neural Network Matrix Factorization (NNMF) [DR15] are two neural approaches for constructing a model that can perform two-way interaction between consumers and items.
2. The integration of MLP with factorization machines form a Deep Factorization Machine (DFM) [Vie18]. The high-order interactions are modeled using the DNN and the low-order relationships among users and items are extracted using the factorization machine.

The working principle of MLP is demonstrated via Fig. 2.2. MLP performs efficient representation of features but it possesses less expressive power as compared to other deep neural networks such as Autoencoder, CNNs, RNNs, etc.

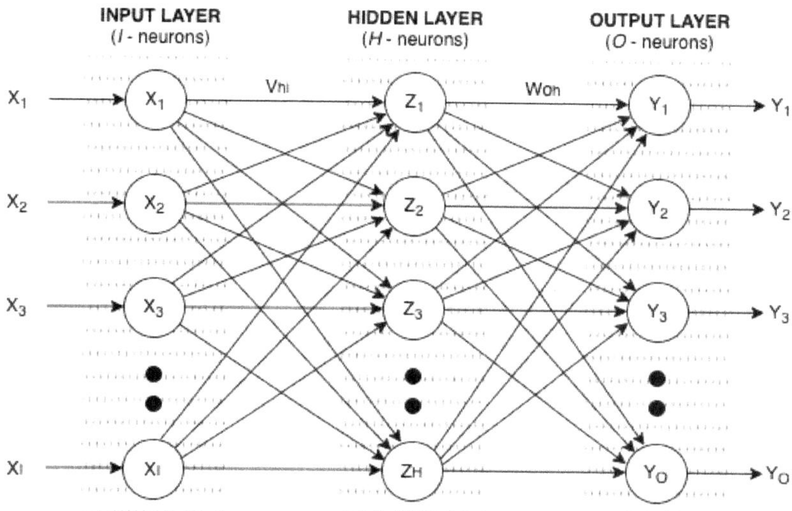

Figure 2.2 Working principle of MLP

1. Wide & Deep Learning for RSs: A single layer perceptron in this model acts as the wide learning component and can also be considered as linear model. It is a satisfying model that can address both classification and regression problems. It was initially developed for Google Play apps recommendation. MLP is used as the deep learning component in this model. The justification for integrating these two learning strategies is that both memorization and generalization are captured by the recommender. Memorization helps in capturing the direct attributes/features from the historical data. This model maintains a balance between the diversity and accuracy of the generated recommendation list. Mathematical representation for the discussed model is given by Eq. 2.1.

$$P(\hat{r}_{ui} = 1|x) = f(Z_{wide}^{T}\{x, \alpha(x)\} + Z_{deep}^{T}\phi^{(l_f)} + B) \quad (2.1)$$

$where,$

$\hat{r}_{ui} => binary\_rating,$

$f(.) => sigmoid\_function,$

$\phi^{(lr)} => final\_layer\_activation\_function,$

$B => Bias,$

$x => input\_feature,$

$\alpha(x) => transformed\_feature,$

$Z^{T}_{wide} => wide\_learning\_parameter,$

$Z^{T}_{deep} => deep\_neural\_component$

[Che+17a] designed a wide & deep locally connected model for performing industry level recommendation by extending the aforementioned concept. The deep learning portion is replaced by an effectively locally linked network, which reduces the operating time by a single magnitude. The transformation of the cross-products must also be designed manually. These preliminary steps are important for the usefulness of this model.

2. MLP for YouTube recommendation: [CAS16] used DNN for generating recommendations in YouTube. The process of recommendation comprised of two basic steps: a) candidate generation and b) candidate ranking. The first step extracts a subset of videos from full collection and the second step ranks the extracted videos based on the neighbourhood computation.
3. MLP for Makeup recommendation: [Ala+17] designed a DNN model in which two equivalent MLP's are included in the research to model labelled scenarios and expert regulations. Both network parameters are concurrently modified by reducing their variations. It demonstrates the utility of the expert information adoption to lead the learning phase in the MLP model. It is very accurate, although the acquisition of expertise requires a lot of human participation.
4. Collaborative Metric Learning (CML): MLP is used in CML to learn representations from object characteristics such as textual data, images and tags [Hsi+17]. The users and items are instructed to optimize the distance between users and

their unwanted artifacts while reducing the difference between users and their desired things.

Autoencoder (AE)

Autoencoder is an unregulated unsupervised model that intends to regenerate the input data at the nodes present at output layer. The mid layer is commonly considered as the most significant attribute describing the input data. Autoencoder is generally referred as a variant of the standard multi-layer perceptron (MLP). AE regenerates input data to better understand the latent functionality of the data by the process of coding and decoding.

AE comprise of a three-layer architecture where the number of neurons present at input layer is equal to the number of neurons present at the output layer. The number of neurons present at the middle layer is much smaller as compared to the number of neurons present at the input layer and output layer. A new signal will be produced at the output layer via the network during training for each training sample. The main aim of network learning is to ensure a similar input and output signal. The reconstruction error denotes this similarity. Through cascading and layered training, the autoencoder will shape a deep structure. Fine tuning can be achieved, after the deep model has been trained via layer-by-layer optimization, by enabling the whole network to recreate the input signal. Following is some of the variants of AE namely: Undercomplete Autoencoders [TA19], Sparse Autoencoders [HZN15], Denoising Autoencoders (DAE) [Vin+08], Contractive Autoencoders (CAE) [ZYX17], Stacked Denoising Autoencoders (SDAE) [MZ20], Deep Autoencoders [CK18], Marginalized Denoising Autoencoder [LKF15], and Variational Autoencoder (VAE) [Ma+19].

Following are the two different approaches for implementing AE in recommender systems:

1. To learn the lower dimension middle-most layer functional representations using the autoencoder.
2. To fill the blank spaces directly in the reconstruction layer of the interaction matrix.

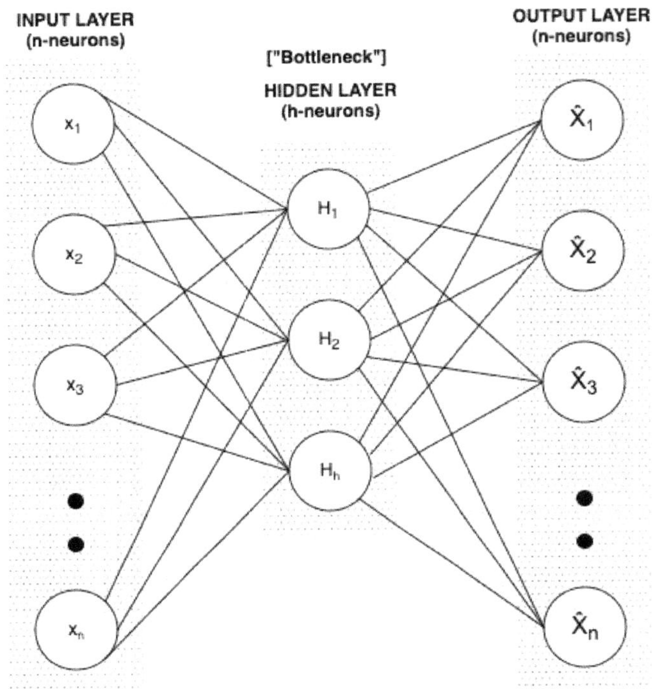

Figure 2.3 Working principle of AE

Collaborative Filtering (CF) based on Autoencoder

AutoRec[Wan+20a] takes r(u) or r(i) component vectors as inputs from the programmer and is intended to recreate them at the output layer. There are two versions reportedly, the AutoRec item-based variant(I-AutoRec)[Che+20] and the AutoRec user-based variant (U-AutoRec)[Che+20], which applies to the two input forms. The mathematical model representation of I-Autorec is given by Equation 2.2 and Equation 2.3.

$$f(R_{*k};\phi) = g(X.h(YR_{*k} + \gamma_1) + \gamma_2) \quad (2.2)$$

$$argmin_{\phi} = \sum_{i=1}^{N} ||R_{*k} - f(R_{*k};\phi)||^2_{\mathcal{O}} + \lambda(||X||^2_F + ||Y||^2_F) \quad (2.3)$$

In the above equation, R k represents the kth column of the utility (user-item rating) matrix. The rating values for unrated items in the utility matrix are kept as '0' by default. The architecture of I-Autorec is defined by Equation 2.2, where, g() and h() denotes the activation functions. X, Y and γ1, γ2 are weight matrices and biases respectively. The output of f() tends to reconstruct the kth column of the utility matrix. In order to minimize the reconstruction error of the I-Autorec, the model aims at minimizing the objective function represented by Equation 2.3. In this equation, $\|\|_{\vartheta}^{2}$ means the output of observed ratings is acknowledged, that is to imply, during back-propagating only weights are modified which are correlated with observed inputs. Following are the four major aspects concerning AutoRec which should be noted before deployment:

1. Since the users' partly measured variation in observed vectors is greater, the performance of I-Autorec tends to act better as compared to U-Autorec.
2. The output of AutoRec is significantly affected by various combinations of activation functions.
3. Boost model efficiency by increasing the number of layers or hidden neurons. This is appropriate since it helps AutoRec to raise the dimensionality of the hidden layer in order to simulate the input functionality.
4. Even more layers can lead to modest improvement in developing a deep network.

CDAE [Collaborative Denoising Auto-encoder][PHY20]: As compared to DeepRec [Liu+20] and AutoRec, CDAE possess following differences:

1. CDAE entry does not have any customer scores, but implicit input 'r' (customer item's preference) is partly observed. If a consumer likes an item 'm', the appropriate entry value is 1, or else 0.
2. CDAE is primarily used for ranking predictions (also termed as recom- mendations of the Top-N preference) whereas AutoRec and DeepRec are mainly used to forecast customer rating scores.

By minimising the overall reconstruction error in Equation 2.4, the CDAE parameters are obtained.

$$\min_{\rho} \frac{1}{N} \sum_{u=1}^{N} E_{p}(r_c|r)[l(r_c, f(r_c))] + \lambda.Reg \quad (2.4)$$

The loss function 'l(.)' can be logistic or square loss. To monitor the complexity of the model, square L2 norm is used by CDAE. The stochastic gradient descent is implemented to learn the parameters of the model and AdaGrad is used to change training phase size automatically during the learning process of CDAE.

Multi-VAE[Multinomial Variational Auto-encoder][Lia+18]: For guidance with implicit information, the Multi-VAE suggested a variational autoencoder, showing better results than the CDAE. While VAEs have been widely researched in image processing and creation, VAEs do not contribute remarkably much study to the domain of recommender systems. Two modifications were made necessary to obtain the most advanced results with VAEs:

1. For the distribution of the data the authors suggested a multinomial probability. They demonstrated that this simple choice achieves models that surpass the Gaussian and logistic probabilities.
2. The traditional VAEs objective, which the authors contend is over- regularized, is reinterpreted and adjusted. They related the learning algorithm that stems from the proposed regulatory framework to the information-bottleneck theory and prejudice toward maximum entropy.

Learning through Autoencoder

Autoencoder is indeed an efficient learning method to feature representations. As such, feature representations of user / item content attributes can also be included in recommending schemes. They are able to learn a complicated representation (manifold) of the data for higher dimensional data, which can explain findings in less dimensionality and is decoded correspondingly into the actual input space. CDL[Collaborative Deep Learning] [Qua+20] derives an accurate profound representation of the items and captures the resemblance and the tacit connectivity between the users and items. CDL is the very first Bayesian hierarchical framework to span the separation between advanced deep learning models and the recommender systems. In addition, CDL can be expanded conveniently to add other auxiliary knowledge to improve the output because of its Bayesian nature.

SDAE [Stacked Denoising Autoencoders] [ZLJ20] is simply a neural network that offers knowledge on the depiction of the input data through understanding how to predict a clean output. Typically, a large bottleneck is restricted to hidden layer in the middle, and the X0 input layer is the noise-corrupted representation of the clean data input. In order to learn SDAE, the regularized optimization problem as shown in Equation 2.5 is solved.

$$min_{X_l, \gamma_l} \|X_c - X_l\|_F^2 + \lambda \sum_l (\|X_l\|_F^2 + \|\gamma_l\|_2^2) \quad (2.5)$$

X_l represents the weight matrix and γ_l denotes the bias vector of the layer 'l'. λ is the regularization hyperparameter. Clean input and noise corrupted input is represented by X_c and X_0 respectively. $||.||_F$ denotes the forbenius norm.

RSDAE [Relational Stacked Denoising Autoencoders] [WSY15] uses a relational information matrix that can be easily incorporated with layered representation learning. This helps the model that is using RSDAE to learn the functional representation from the knowledge about the contents and the relations between items simultaneously. CVAE [Collaborative Variational Autoencoder] [LS17] is fundamentally a generative latent model of the variable, where the contents of the products are triggered by their variable latent contents and review scores are triggered by latent items. Latent item variables include both information on content through latent contents and collaborative relevant data via the collaborative latent variables. By using variational autoencoder, CVAE replaces the collaborative deep learning.

CDR [Collaborative Deep Ranking] [Yin+16] incorporates abstract delineation of side details of items into the pair-wise ranking model's Bayesian structure. The feature delineation of the contents of the item is derived from SDAE directly. Collaborating with SDAE, the pair-wise ranking factor can cope to some degree with sparsity problems and can increase the consistency of the recommendation. Although CDR uses SDAE to represent features, CDR can also work collaboratively as a generic framework with several other deep learning techniques, including CNNs and RNNs. Experimental results shows that the performance of CDR is better than collaborative deep learning in case of ranking predictions.

AutoSVD++ [ZYX17]: By grouping users who give the same tacit input together, AutoSVD++ dramatically boosts the efficiency of computation. AutoSVD++ combines contractive autoencoder with SVD to enhance the performance of recommender systems. By incorporating the contractive autoencoder, the recommendation model takes into account the nontrivial and nonlinearity of item content information and learn the semanticized representations in a restricted integration space. AutoSVD++ possess several advantages such as:

1. Contractive autoencoder conveys insignificantly small input variations in comparison to several other variants of autoencoders.
2. In order to increase the precision value, AutoSVD++ models the inferred feedback.
3. An appropriate algorithm for training is suggested to minimize the training period.

HRCD [Hybrid recommendation model with CF and deep learning] [Wei+16] was designed to address the problem of cold start in recommender systems. It analyses and applies the item characteristics obtained from the deep learning neural network to the timeSVD++ CF model. It is a time-conscious model utilising SDAE to know and understand items from raw characteristics and aims to solve the problem of cold start.

CNN [Convolutional Neural Network]

CNN [ZYX17] comprise of convolution layers and bundling processes. It captures global and local characteristics and thereby vastly increases performance and precision of the model. The unstructured multimedia data processing is very efficient in case of CNN models. The working principle of CNN is illustrated by Fig. 2.4.

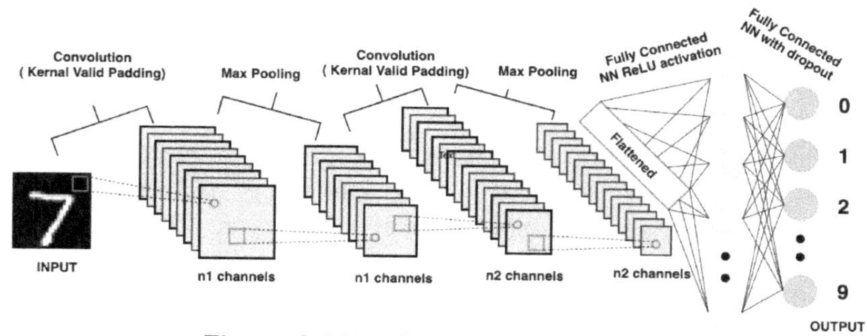

Figure 2.4 Working principle of CNN

CNN [ZYX17] comprise of convolution layers and bundling processes. It captures global and local characteristics and thereby vastly increases performance and precision of the model. The unstructured multimedia data processing is very efficient in case of CNN models. The working principle of CNN is illustrated by Fig. 2.4.

CNN for Feature Extraction

CNNs may be used to reflect features that are learned from various sources such as text, images, audio, video, etc.

[Wan+17b] discusses the effects of visual characteristics on the recommend- ation of point of interest and introduces an improved POI recommendation framework for visual information. This method adopts a CNN method to derive image features dependent on PMF in order to investigate the relationships between visual content and the latent factors (location / user). In general , four facets are based on the current POI recommendations: spatial impact, societal correlations, temporal trends and indicator of textual contents. These facets have proved effective in enhancing point of interest recommendations.

[Lei+16] builds a deep dual-network in which the two subnetwork maps users' input images and interests into an alike latent semantic space, and then it performs the process of decision-making by measuring the distances between the user and the input image present in the same latent space. The approach of comparative deep-learning is used to shape the deep-network utilizing a pair of images to understand the trend of their relative distances compared to one individual. The network comprises of 2 CNNs for understanding image representations and an MLP for modelling consumer expectations.

[RK16] designed a CNN-NN framework for addressing the problem of multi- label classification. The network is composed of the CNN and NN components. The CNN part derives the graphic details from the image pixels and models the user context in the NN [Neural Network] component. The outputs of two NNs are merged and fed into Softmax function to estimate the likelihood of nominee tags.

[ZNY17] constructed a DeepCoNN[Deep Cooperative Neural Networks] that comprise of two parallel NNs merged in the last layers of the network. One of it's networks relies on knowing consumer actions by utilizing customer feedback and the other on the properties of the items published in the reports. A generic (common) layer is established on the top for pairing these two NNs. The generic layer helps users and items to communicate like factorization machine techniques on the acquired latent factors. In order to deaden the problem of sparsity in recommender system, the DeepCoNN model leverages the rich semantic features of the review texts. This ultimately enhances the interpretability of the DeepCoNN model.

[She+16] proposed a model that can be used to extract features from the text information. They built an automatic resource recommendation model based on CNN. The CNN plays a key role in extracting the item features from the available text information such as content of the course material, introduction of the text, etc. The suggested approach explicitly controls the quality of the learning tools without human tagging. The outcome of the experiments performed on the proposed model reveals that CNN is a successful model to bridge the semanticized distance between texts and the latent vectors.

[VDS13] targeted to solve the problem of cold start in music recommendation. They explored the usage of CNN to model music audio latent factors if they are not obtainable from data used. Although audio signals do not anticipate any of the song's characteristics that influence consumer behaviour, the resulting suggestions are responsive. The prediction of latent factors is a viable way to advise new and unpopular music.

[Lee+18] suggested a deep neural network that integrates videos into a metrical space that maintains video-to-video relationships through their audio- visual information.

Collaborative Filtering based on CNNs

It is feasible to use CNNs explicitly to work together with vanilla CF based recommendation models. [He+18b] designed a new multi-layer NN architecture to leverage the properties of collaborative filtering in recommender system. The proposed ConvNCF architecture is used to specifically model the relationships between the parameters of the embedded space by utilising an external component. CNN is used in the model to learn the higher-order correlations among the parameters of the embedded space. [TW18] introduced sequential recommendations with the CNN's where the sequence-awareness guidance is focused on a hierarchical and vertical CNN for creation of union-level sequential trends. The concept is to incorporate series of recent items into some kind of "image" in times and latent spaces by means of convolutionary filters and to learn sequence patterns as the local features.

Recommendation based on Graph CNN

In recommendation activities, graph CNN could easily control the interactions. Graph CNNs acts as a useful resource for platforms that work on non-Euclidean data such as: networks of protein interactions, social networks, etc. [BKW17] saw the issue of recommendation as a challenge of graphical CNN forecasts. The proposed system enabled user-side knowledge such as social networks and item interaction information in the recommendation model to be implemented quickly. [Yin+18] model produces items embedded from both the graph layout and item features using graph CNNs and random walk, and thus blends in well with a wider network of recommender system.

RNN [Recurrent Neural Network]

RNNs give traditional neural networks an intriguing twist. In circumstances with a 'sequence' style entry with no standard in size, a neural network of vanilla takes a vector of a fixed size as an input, which restricts its use. RNNs have been configured to accommodate a range of inputs that have no predetermined size limit. RNN is well adapted for sequential data propagation. It recalls past computations using loops and memories. The output(s) may be determined not only by limitations implemented on inputs perhaps with a regular neural network, but also through a "hidden" state vector reflecting context that is centred on the previous input / output(s). So, based on earlier input in the sequence, the same input could generate different results. In session-based suggestion functions, RNNs can cope with the time complexity of interactions and asynchronous user behaviour. The working principle of RNN is illustrated by Fig. 2.5.

The framework typically would not distress people to log into certain real- world apps or websites in order not to view their long-term patterns of usage or long-term ambitions. However, these structures enable users to gain short- term expectations via the cookie or session mechanisms. [Hid+15] proposed a recommendation model centred on session, where current status of session is taken as input with 1-of-N(items) encoding. If the respective item is triggered in that session, the coordinate would be 1 otherwise it will be 0. The outcome represents the chance of each item to be next in the session. [Wu+16b] offers a session-based e-commerce platform suggestion interface. It uses the key RNNs to determine the click history and analyzes what consumer may purchase next. It just retains a small number of recent states while the older states fall into one historical state in order to mitigate their computation costs. This way, the model helps in maintaining the balance between accuracy and computation cost. [Wu+17] foresees the possible behavioural direction. This is accomplished by using LSTM for both users and items present in the system. RNN also seems to be a good alternative for gaining side knowledge using sequential layout. [Dai+16] provides a latent co-evolutionary paradigm to capture latent characteristics of users and items in a co-evolutionary fashion. In influencing adjustments to consumer desires and items status, the connections of users to items play a major role. The author suggested using the RNNs to model the historical experiences to automatically grasp the effect of drift, progression and co-evolution of consumer and item characteristics. [BBM16] suggested text series to be stored in a latent factor pattern utilizing GRUs. This hybrid approach explores challenges with both warm and cold start. In addition, the authors have implemented a regularized multi-task to avoid overfitting and mitigate training data sparsity. The primary purpose is to determine the rating scores and the secondary role is the items meta-data prediction. [Oku+17] recommends the usage of GRUs to acquire more description about the past of user surfing and to suggest news items with a model of latent factor. Comparing with the conventional terms strategies, the findings are greatly better. This technology has completely been introduced to support over 10 million unique users per day for online development services.

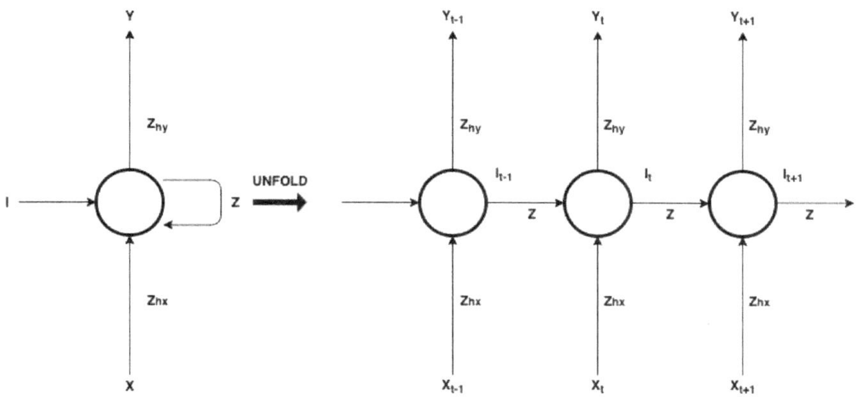

Figure 2.5 Working principle of RNN

RBM [Restricted Boltzmann Machine]

RBM is a graphical(undirected) model that has played a significant part in the recent Deep Learning Systems. It comprises of two layers namely: visible and hidden layer. The [DBN]Deep Belief Network initially pre-trains the weights of the network layer by the RBM and afterwards utilizes the gradient approach to adjust these weights. It should also be noted that RBM & AE are often used in deep learning as the fundamental building blocks. Unlike the AE, however, the RBM seems to be the energy-based model and can also be referred as a specific form of Markov Random Field. The general RBM is considered as an effective approach for representing dependencies between the random variables. The term 'restricted' in RBM means that there is no communication (intra-layer) in visible or the hidden layers. Each node is a computing locus that processes information and progresses by taking stochastic assumptions on whether to send that information. The working principle of RBM is illustrated by Fig. 2.6. The training objective is to decide how well these different pairs of variables are effectively related. By looking at Fig. 2.6, it is clear that RBM possess two basic properties:

1. On the forward pass, with the help of hidden biases, the activations are produced by the RBM.
2. On the backward pass, with the help of visible layer biases, the reconstruction is learnt by the RBM.

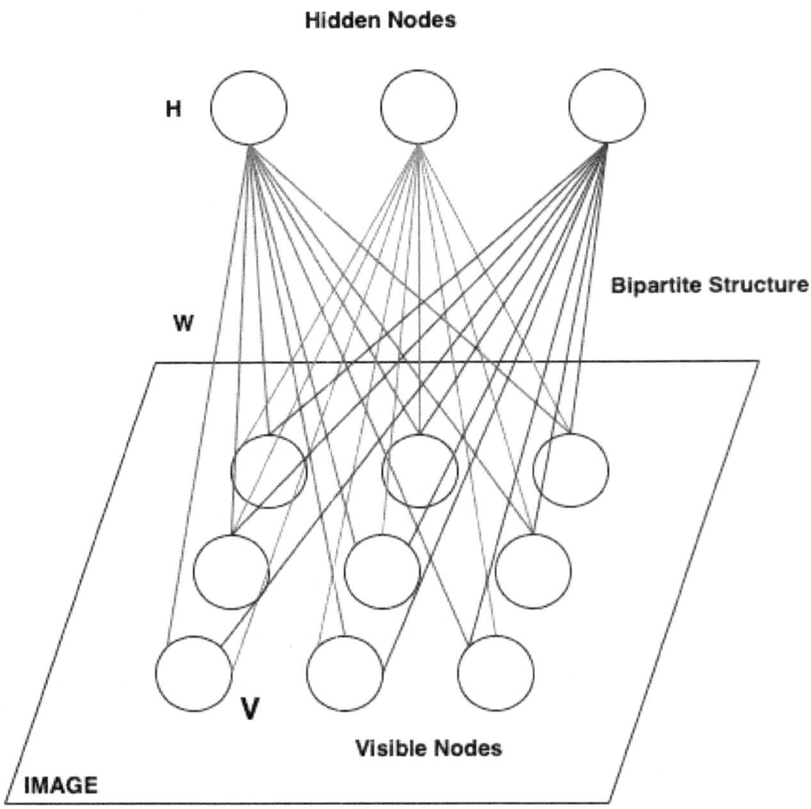

Figure 2.6 Working principle of RBM

Equation 2.6 defines the energy function of RBM.

$$E(V, H) = -a'V - b'H - H'WV \quad (2.6)$$

where, a and b represent the offset of visible and hidden layer respectively. 'W'. denotes the weight value that connects the hidden and visible layer in a Restricted Boltzmann Machine. Equation 2.7 delineates the free energy formula for RBM.

$$F(V) = -a'V - \sum_{i} \log \sum_{H_i} e^{H_i(b_i + W_i V)} \tag{2.7}$$

Because of RBMs unique nature, visible and hidden components are significantly different with each other. So, we can represent it as in Equation 2.8.

$$\begin{aligned} P(H|V) &= \Pi_i P(H_i|V) \\ P(V|H) &= \Pi_j P(V_j|H) \end{aligned} \tag{2.8}$$

Collaborative filtering based on RBM

[SMH07] constructed the first recommender system that was based on RBM. RBMs visible unit is restricted to binary values, so the rating values are expressed in a one-hot vector format to adjust for this constraint. Each consumer seems to have a unique, shared parameter RBM. Contrastive Divergence approach is used to learn the parameters of the constructed RBM. The essence is that by providing scores, users indirectly reveal their interests, irrespective of how they rank items.

[GN13] incorporated user and item-based RBM-CF in a single system. In the proposed system, the visible units are decided by hidden units of both users as well as items. They found that by using a real-valued visible layer that explicitly models user-item ratings is perhaps more effective and easier, and ultimately yields improved predictive efficiency while using a multinomial, visible binary layer. This extends to all except standalone user-based or item-based RBM models, in which the multinomial model yields stronger performance after several iterations (epochs).

[Liu+15b] designed a hybrid recommendation model based on RBM. The constructed RBM-CF model incorporated features of the items and is based on conditional RBM. The conditional layer here is modelled on binary information of item genres, impacting both the hidden layer as well as the visible layer with distinctly connected weights.

Neural Attention Models

Neural attention models are variational neural architectures which work over an input series or input picture dependent on soft content. They are driven by human sensory interest and can strip out uninformative features from raw sources and eliminate the adverse effects of noisy data. This system of attention is omnipresent and yet successful in several fields such as NLP[Natural Language Processing] and Computer Vision. Somewhere within recommendation frameworks, the attention procedure may be used to sort out the noisy information and pick the most relevant products while maintaining strong interpretability. Attention paradigm learns to be attentive to input ratings. In the heart of the model, attention values are determined. The neural attention framework is divided into two groups, depending on the way of determining attention scores. They are as follows:

1. *Standard Vanilla Attention:* It uses a vector with fixed parameterized context for learning to attend.
2. *Co-Attention:* It uses two-sequences for learning the attention weights. Self-attention is a special Co-attention case.

[CJ18] uses a diligent collective filtering model with a 2-level attention mechanism within the latent factor model. The model consists of the element level and the component level: where the element level one chooses the most representative elements for characterizing individuals, and the component level one collects the most informative details for any individual.

[Li+16] constructed an LSTM model based on attentions for recommending hashtags. The model uses both RNNs and a sequential property control system to identify insightful terms from the microblog messages.

[GZ16] designed a CNN model based on attention for the same microblog hashtag recommendation that is viewed as a multi-label classification issue. The model comprises of a local and global attention channel. The local channel has a window-size and threshold attention layer for selecting insightful terms whereas the global channel has a converging and max-pooling strata for all the terms.

NADE [Neural AutoRegressive Density Estimation]

NADE was introduced to address the challenge of density estimation and unsupervised distribution. They use the probability product concept and a weight-sharing method influenced by RBM to generate an estimator that can be traced and generalized well. The working principle of DeepNADE is illustrated via Figure 2.7. [Zhe+16] proposed CF-NADE (Collaborative filtering based on neural autoregressive density estimation) which models the rating score distribution. CF-NADE uses chain rule to model the joint probability as discussed in Equation 2.9.

$$P(R) = \Pi_{i=1}^{Z} P(R_{m_{o_i}} | R_{m_{o_{<i}}}) \tag{2.9}$$

where, Z represents the total items rated by the user, 'o' is the Z-Tuple, R_{moi} is the rating score for item m_{oi}.
Suppose the user has rated four movies i.e., m1, m2, m3 and m4 with score value of 3, 1, 2 and 4 respectively. CF-NADE follows following procedural steps:

1. What is the probability of user rating the movie m1 with 3 stars; no conditions defined.
2. What is the probability of user rating the movie m2 with 1 star; given condition that movie m1 is rated with 4 stars.
3. What is the probability of user rating the movie m3 with 2 stars; given condition that movie m2 is rated with 1 star and movie m1 with 3 stars.
4. What is the probability of user rating the movie m4 with 4 stars; given condition that movie m3 is rated with 2-star, movie m2 is rated with 1 star and movie m1 is rated with 3 stars.

The sequence of the movies should preferably adopt the timestamps of the ratings. Empirical evidence indicates, however, that random drawing often provides successful outcomes. This concept may be generalized further to construct a deep model.

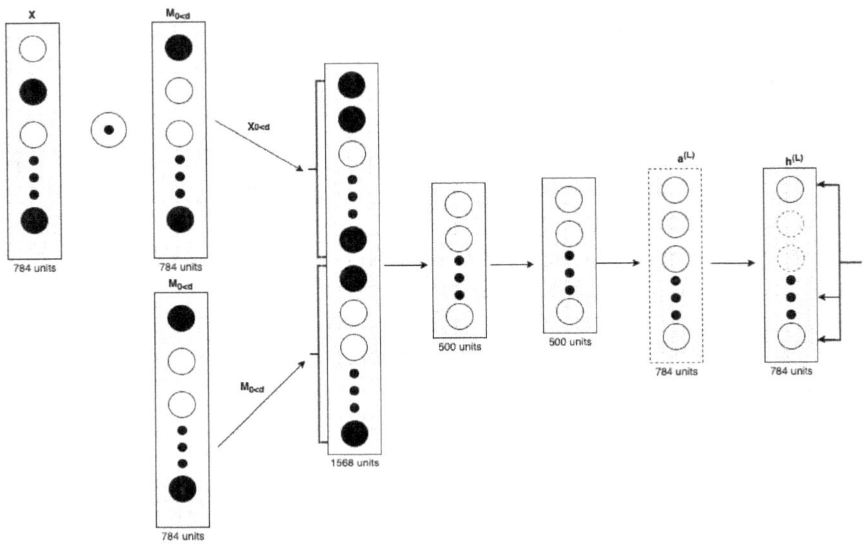

Figure 2.7 Working principle of DeepNADE

DRL [Deep Reinforcement Learning]

RL [Reinforcement Learning] works on the principle of trial and error. It comprises of five components namely: actions, agents, environment, reward and state. DRL formulates the combination of DNN and RL which accomplished human level efficiency in diverse disciplines including games and self-driving vehicles. DNN empowers the agent to gain knowledge of raw data without manual controls and domain heuristics, and to extract reliable representations. The learning cycle of reinforcement learning is illustrated by Figure 2.8. The process starts with the environment getting observed by the agent (STEP-1) and obtains a reward and state. The agent uses the obtained state to decide the next phase (STEP-2). The agent subsequently sends an action to the environment to try and manage it in a desirable manner (STEP-3). Finally, as a product of the agent's actions and previous state, the environmental transformations and their internal status adjusts (Step 4). Then, the process (STEP-1, STEP-2, STEP- 3 and STEP-4) repeats. The recommendation is traditionally considered as static in most recommendation models, making it difficult to grab the temporal willingness of the user and respond promptly. DRL has rendered use of the customized recommendation in recent years.

[Zha+18b] proposed a deep learning-based recommender system that allows its strategies to be constantly improved during its encounters with users. They used Reinforcement Learning (RL) to simplify the learning of optimal strategies by suggesting test and failure items and confirming them with user input.
Feedback from consumers can be both positive and negative. Both feedback forms can enhance the power of constructed recommender model.

[Zha+18a] constructed DRL based recommender system (DeepPage) for page-wise recommendation. They addressed two problems in their model: (a) How to update the recommendation strategy based on user feedback in real time. (b) How to obtain a page with good display things that are enormous challenges for conventional suggestion systems.

[Zhe+18] proposed a system using DRL to detect complex shifts in news coverage and customer preference, integrate customer return trends and improve suggestion variety.

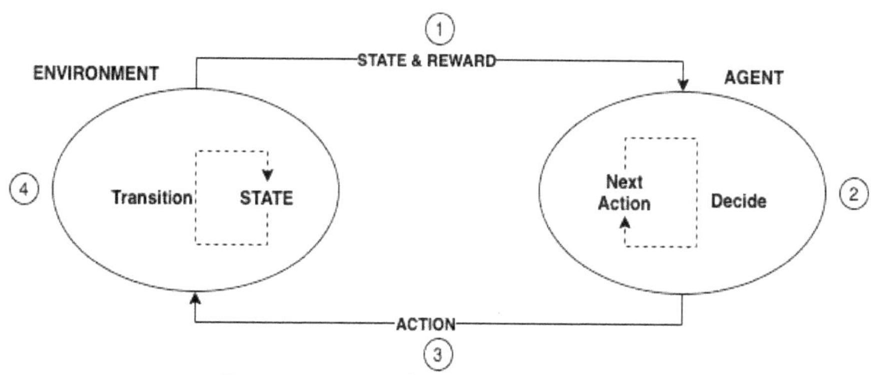

Figure 2.8 Working principle of RL

Adversarial Networks

Adversarial networks comprise of two principle components namely: generator and a discriminator. These two NNs are concurrently trained by engaging in a minimax game. There are possibly two schools of thought in retrieval of knowledge, generative and discriminatory retrieval. Generative retrieval presupposes that there is a generative mechanism behind document and query, and the retrieval can be done by producing appropriate document 'D' for any given query 'Q'. Discriminatory retrieval will estimate significance score 'r' provided the appropriate pairs of query-document records. IRGAN [Wan+17a] goal is to merge these two concepts into a single paradigm and to make them play a minimax game like GAN generator and discriminator. The objective function for the IRGAN model is represented by Eq. 2.10.

$$\min_{\alpha}\max_{\beta}\sum_{i=1}^{n}\left(E_{D}\ u_{T(D|Q_{i,r})}[\log Z(D|Q_{i})] + E_{D}\ u_{\alpha(D|Q_{i,r})}[\log(1 - z(D|Q_{i}))]\right) \tag{2.10}$$

where, Z(D|Qi) = σ(fβ(Q, D)), σ denotes the activation function, α and β represents the generative and discriminative retrieval parameter respectively. These two parameters are learned via approach of gradient descent.

$DU_T(D|Qi,r)$ reflects the preference distribution of the user. The generative model D Uα(D|Qi,r) tries to estimate the true distribution and the discriminative model fβ(Q, D) attempts to discriminate between relevant and irrelevant documents.

[He+18a] suggested an adverse custom rating method that boosts Bayesian customised training ranks. It plays minimax game between the initial BPR goal and the adversary that introduces sound or permutations to increase the BPR loss.

Deep Hybrid Models

Numerous neural building blocks can be combined with the excellent versatility of deep neural networks to formalise more efficient and descriptive models. It is recommended that the hybrid approach should be moderately and thoughtfully constructed for the assigned purposes, despite the massive possible forms of combination. The combination of (CNN and AE) [Zha+16], (CNN and RNN) [Zha+17], (RNN and AE) [WXY16], (RNN and DRL)[Wan+18b] are some of the deep hybrid approaches for recommendation that has achieved wide spread success in several application domains. [Zha+16] integrates CNNs and autoencoders to retrieve image features. It effectively uses embedding strategies to use structural information, textual content and visual information. [Lee+16] proposed a combination RNN and CNN model that proposes quotes, creating a rated collection of quotes that are implied by the demand text or dialogues. The CNNs are used to learn the relevant local semantics via messages and transform them to a representation vector. Then LSTM processes these vectors to evaluate the importance of destination quotes in the individual tweet dialogues. [Che+17a] combined CNNs and RNNs for custom keyframe suggestion into videos, in which CNNs acquire feature vectors via keyframe objects and RNNs are used in textual processing. [EF17] incorporated CNNs and RNNs into a citation suggestion encoder-decoder system. The encoder CNNs catches the long-term addictions of the citation background, whereas the decoder RNNs recognises the chance of a word in the title of the cited document in all previous terms coupled with representations acquired from the CNNs. [WXY16] uses RNNs & denoising autoencoders to address shortcomings, including insufficient robustness and the failure to model text sequences. They designed and presented the hierarchy of the Bayesian model of recommendation called CRAE, a generalisation for RNNs. The proposed model is composed of encoding and decoding components which uses feedback-based neural layers with RNNs to collect sequential content data. For a medication recommendation [Che+17b] incorporates the supervised profound affirmation preparation and RNNs.

2.4 Research Direction

The existing DL strategies significantly contribute to the building of a solid base for the recommender framework. Every DL model has its own specificity for data processing. CNNs, for example, can be viewed as the optimal solution for the content-based functional learning tasks and can easily resolve the issue of cold start. The RNNs are indeed very effective in many different activities such as multi-modal training, pattern forecasting and recommendations based on sequential data. The different forms of DL approaches improve the recommender efficiency by making valuable improvements on the standard recommendation system, which opens up a new outlook on the deep recommendation system. The recommendation focused on DL is therefore still emerging and new issues are being faced by evolving research developments. Hereby we have presented some of the emerging research trends with respect to deep recommender systems.

1. Addressing the new evaluation metrics: Researchers found that concentrating on enhancing the precision/recall or diminishing the estimation error is not enough for building trust on a recommender framework. Several other parameters such as: Diversity, Novelty, Serendipity, Coverage (Item/User/Catalog), privacy, interpretability, etc should be taken into consideration for evaluating the deep recommender systems.
2. Scalability: The scalability challenge concerns the rise in the amount of data for products and consumers and their embedding. Scalability is crucial in selecting the recommendation model, where the time is also significant consideration. A recommendation method should concurrently satisfy the two considerations. DL in this regard justifies its strength in processing the big data. Further future studies on how to suggest accurately addressing the following issues should be analysed: (a) Incremental training and learning for streaming data such as large number of incoming consumers/products, (b) Efficiency computation for digital media information sources, (c) Model simplicity and interoperability balance with the increasing parameters.
3. Bilateral learning of consumer content and product features: Precise suggestions require a thorough understanding of the features, the actual needs and preferences of the consumer. This may, of course, be done by the use of ample auxiliary knowledge. For instance, context information adapts services and items according to the circumstances and environments of the consumer and mitigates the impact of cold starts. The potential to understand deeper in the analysis of heterogeneous data sources provides more possibilities to propose different items with unstructured data like audio and videos. Feature engineering in the recommendations research group has still not been thoroughly explored, but it is important and widely used in different industrial domains.

Most of the current models entail time-consuming, complicated crafted and specified features. DNN is a viable approach for enduring features automatically. Further research on deep feature engineering unique to recommending systems could save human time and enhance the consistency of recommendations.

4. Intelligible deep learning model for efficient recommendation: DNN are highly uninterpretable as an obvious description. As such, it seems like producing explanatory recommendations is an uphill job. In the same way, it would therefore be reasonable to believe that massive, complicated neural models simply make the data understandable. There are two key aspects that explains the importance of DL. The first is to give users simple forecasts that enable them to consider the reasons behind the recommendations of the network. The second track focuses mainly on explaining the individual, testing weights and events to learn much more about framework. To this end, it will be a positive direction and next step to build better cautious frameworks, ideally at the level of communicative or generative reasons. Since models are now able to demonstrate what leads to the judgement, we assume this will be the next limitation.
5. Problem solving for recommendation: The next limit for recommenders is maybe to respond to circumstances involving multi-step deduction and justification. Example: explaining the social profile of a user and purchases etc., thinking about multiple ways of recommending a product. In summary, we should anticipate the logical architectures to begin proposing machine analysis in the first place.
6. Cross-domain Architecture: The single domain recommendation frame- work only concentrates on one domain, thus dismissing consumer curiosity on other domains. A cross-domain recommendation framework that follows the goal domain recommendation through learning from source domains gives the optimal answer to these problems. Transfer learning, which is structured to enhance learning activities in one domain by utilising information passed from other domains, is amongst the most discussed topics in the cross-domain architecture.

7. Multi-task learning: Threefold benefits of using DNN based multi-task learning are: (a) Learning multiple tasks concurrently will evade overfitting by making generalisations on the hidden interpretations, (b) The auxiliary task offers a concise contribution to justify the recommendation, (c) Multiple tasks include significant increase in data to mitigate the issue of sparsity. In addition to implementing side tasks, multitasks of cross-domain learning may also be used with each task to produce a recommendation for every domain.

2.5 Summary

In this chapter, we analysed our assertions and observations related to DNN based recommender systems. The reader is offered with several perspectives based on the overall study of DNN based systems. As we addressed in depth the existence of DL techniques in many aspects of computer science, these approaches display reliable and successful success in recommender frameworks. The incorporation of numerous data sources in enhancing the machine capability, the storage and effective analysis of large data are prominent explanations of contemporary supremacy in DL based applications.

Deep learning methods are not trained in a particular form of recommendation; they are used in all forms of methodologies for suggestions for multiple purposes. In CBF, these approaches are often used to draw features from heterogeneous information sources to create content-based (user / item) profiles. In CF however, they are also used to remove latent factors from the user's matrix as a model-based method. In hybrid systems, deep learning approaches are used to derive and incorporate functionality from auxiliary information. Amongst all DL methods, latent factor research involves RBMs, AE and NADE. Although current studies indicate that NADE supersedes RBM and AE-based recommendation systems in terms of accuracy. AEs are more common for building the recommender system because of their easy construction, feature engineering, reduction in size of data and lack of value estimation capabilities. Existing literature indicates that DL methods are more effective than standard existing algorithms like matrix factorization and neighbouring methods. The key explanation is that DL algorithms provide non-linear interpretations of user interests that enable to detect unforeseen or nonsensical behaviour of the users. We mention insufficiencies that future researchers should concentrate on while building a recommender system. (a) Extra details need to be incorporated into consumer preferences to boost performance and manage problems in existing recommendation systems. In e-commerce, effective recommendations may be made using the contextual info, tags and consumer expectations in connection to the user-item interest matrix. Thus, DL techniques may be used more compositely to derive valuable knowledge from various sources of data and to incorporate it into the recommender system. (b) Deep learning methods gain high precision when used with non-privately acquired interests in the generated recommendations. It is a task to use DL techniques on privately gathered data to balance competing priorities of precision and privacy thresholds in recommendation systems. Moreover, using such techniques in private collections, either centralised or distributed, is yet another challenge. (c) Deep learning is effective in extracting characteristics. It was, however, not used to identify closest neighbours of active accounts in memory-based CF that may further boost their precision. DL approaches to identify successful consumer neighbours to increase precision is still a major challenge.

CHAPTER 3

Recommendation with Deep Learning Techniques in Content Based System

"Content builds relationships. Relationships are build on trust. Trust drives revenue" — Andrew Davis

The content-driven engines were precisely based on a simple ideology: "The set of items similar to that the user enjoyed in the past will act as engrossing items in future". This approach is designed by discovering patterns that compare the key factors that influence items that are stored in the user's profile that he/she liked, with those that describe the items which the user has still not reviewed. The significance of learning an effective assessment of the items is instantly carried out by such a paradigm: the more correct the interpretation, the better the user can receive recommendations. In recent years, the approaches of deep learning has proved to be exceptionally valuable in several domains of machine learning. In this chapter, we present a Hierarchical Autoencoder Content-based Recommender System (HA-CBRS). In the proposed hierarchical model, every layer is reconstructed; complementary knowledge is given by each layer. In order to produce the final result, the output from all of the layers is blended. Authenticity and performance of the proposed model is tested on MovieLens-100K, MovieLens-1M, MovieLens-10M dataset. The superiority of the model is exhibited by the high accuracy obtained by the proposed model as compared to numerous other state-of-art comparative models.

3.1 Introduction

The main objective of a personalization system is to orient the users towards interesting or relevant items from a large pool of possible alternatives in a personalized way. Typically, such systems obtain better understanding of the needs, preferences and interests of user groups and adjust their behaviour on the basis of such info, by assisting people throughout many decision-making activities. Recommendation mechanisms can help users overcome the issues of information overload since they can prioritize content as per user needs. To recommend items relative to some of those a particular user has previously enjoyed, content-based recommendation systems ought to accurately portray user profiles and items.

Deep learning (DL) methods have gained prominence as they have demonstrated a very strong performance in a broad spectrum of applications. A very concise interpretations that DL techniques are capable of learning by trying to exploit neural networks are indeed directly attributable to such efficiency. Bidirectional RNNs [Mus+18] have started to emerge as among the most efficient interpretations in this sort of situation, as they are very decent at modelling data sequences and can gain knowledge relationships with prior and future information as well. Since it can be worked out as a series of words, such a function makes them very convenient for modelling textual information as well. The utilization of information graphs [BHB11] that will further feed the depiction with new and descriptive functionality is another fascinating development in the field of content-based RSs.

The training techniques remain surprisingly simplistic, in accordance with the increasing complexity of neural networks. The stochastic algorithm is the basis of the most functional optimization techniques for DNNs. Nevertheless, the learning parameter of SGD is sometimes difficult to adjust. Since various parameters magnitude differ dramatically, it is utmost necessary that the training phase should be adjusted throughout the process. Numerous adaptive SGD's variants have been designed to address this issue. Algorithms such as Adam [Wan+19], Adagrad [Zou+19], RMSProp [Zou+19] are meant to dynamically tailor the learning rate dependent on gradient statistics to various parameters. While they generally optimize the configuration of learning rates and contribute to faster convergence, but in some cases their reliability of generalisation is considerably weaker than that of SGD. In our proposed framework, we have used ND-Adam [Zha18] which specifically regulates the step size and direction of updation in a very meticulous way. ND-Adam can achieve considerably decent generalization improvement over Adam optimizer.

In this chapter, we present the user preference using deep hierarchical autoencoder and features gathered from the items. The preference of the user can be collected using explicit or implicit data. Collecting implicit scores is easier as compared to explicit ones. It

should be mentioned, though, that gathering negative reviews for implicit scores is almost impossible. If an individual has not invested time on a specific thing, one would not understand that this is because he/she doesn't like it or that he/she doesn't know about it. We would therefore limit ourselves to explicit scores in this chapter.
In brief, we offered the below-mentioned contributions in this chapter:

1. Use of Hierarchical Autoencoder by maintaining the balance between decoder and encoder level for effective representation of explicit rating values.
2. Administrating the training of HA-CBRS using ND-Adam optimizer.
3. Deciding the final order of recommendation by using item attributes (genre, release year) as reference data.
4. Justifying the proposed model performance using metrics such as MAE, RMSE, Precision and Recall.

The remaining section of the chapter is structured as follows: Section 2 demonstrates the techniques used by the proposed HA-CBRS architecture. Section 3 discusses the flow model of proposed architecture along with highlighting of major contributions. In Section 4, we outlined the experimental outcomes obtained at each step of the proposed model. We also illustrated the performance of the model using different performance metrics such as MAE, RMSE, Precision and Recall. Section 5 contrasts the model reliability and efficiency with respect to numerous other state-of-art personalization models.

3.2 Experimental Methodology

In this section, we have discussed the different techniques clubbed together to generate an appropriate order of recommendation.

Hierarchical Autoencoder

Autoencoders seem to be an unregulated process in where we use neural networks to learn representation. In particular we can build a neural network architecture in order to enforce a network bottleneck that causes the original input to be interpreted by a compressed information. This compression and further rebuilding would be a very challenging job if the input characteristics would be independent. However, if there is a certain kind of structure in the data (i.e., similarities amongst attribute values), it can also learn and thereby elevate this structure as it pushes input via the breakdown of the network. Autoencoders comprise of two components namely: encoder and decoder. Encoder is responsible for performing mapping of input values to the representation, whereas decoder maps the representation to output values. An ideal autoencoder seeks to maintain the balance between following:

1. Input receptive enough to create a reconstruction correctly.
2. The inputs are dismissive enough so that the model does not merely contain or overfit the training phase.

This balance requires the model to attain only the adjustments in the data necessary to redefine the input, without securing redundant information in the input. This includes in perhaps the most cases creating a loss function, with one term promoting a model that is adaptive to inputs as well as the second term restricts overfitting. The aforementioned trade-off is represented by Equation 3.1

$$L(X, \hat{X}) + R \tag{3.1}$$

where, X and $\hat{X}$ denotes the value at input and output layer respectively. 'R' represents the regularization term.

The input side of autoencoder is fed with rating values which are ordered either based on users/items. The same evidence of information is displayed at the output side. As the input seems to be incomplete/missing, the learning of the autoencoder has to be marginally modified; it cannot be viewed as empty. Instead, where a missing value happens, the association will not be adjusted by backpropagation – essentially nullifying the effects of the missing value. The inputs are yet again introduced to the autoencoder after the training is finished and the learned weights are introduced for imputation. By dropping the information at input and not from the output inclines the system towards denoising autoencoder. The autoencoder here utilizes a predominating representation that boosts the performance.

Our proposed architecture is based on the formulation of asymmetric autoencoder that has only one decoder. The main advantage of this asymmetric encoder is, unnecessarily we need not to train the different layers of decoders. The acquisition of much more parameters results in overfitting which declines generalisation and affects performance. The mathematical model for asymmetric formulation of stacked autoencoder is represented by Equation 3.2.

$$\begin{aligned} T_x = Z_D \alpha(Z_{E4}\alpha(Z_{E3}\alpha(Z_{E2}\alpha(Z_{E1}T_x)))) \\ Z_{E1}, Z_{E2}, Z_{E3}, Z_{E4} \in 4 \text{ Encoders} \\ Z_D \in 1 \text{ Decoder} \\ T_x \in \text{Training data} \end{aligned} \tag{3.2}$$

During the training phase, the weight values are learnt by significantly minimizing the Euclidean norm. The mathematical formulation for this minimization is given by Equation 3.3.

$$\min_{Z_{E4},Z_{E3},Z_{E2},Z_{E1},Z_D} \|T_x - Z_D\alpha(Z_{E4}\alpha(Z_{E3}\alpha(Z_{E2}\alpha(Z_{E1}T_x))))\|_2^F \tag{3.3}$$

In our proposed architecture, we performed slight modification in the above discussed stacked autoencoder. At every encoder level, we proposed to learn a decoder. Mathematically, it can be demonstrated via Equation 3.4.

$$\begin{aligned} \min_{Z_{E4},Z_{E3},Z_{E2},Z_{E1},Z_{D4},Z_{D3},Z_{D2},Z_{D1}} \\ \|T_x - Z_{D4}\alpha(Z_{E4}\alpha(Z_{E3}\alpha(Z_{E2}\alpha(Z_{E1}T_x))))\|_2^F \\ +\|T_x - Z_{D3}\alpha(Z_{E3}\alpha(Z_{E2}\alpha(Z_{E1}T_x)))\|_2^F \\ +\|T_x - Z_{D2}\alpha(Z_{E2}\alpha(Z_{E1}T_x))\|_2^F \\ +\|T_x - Z_{D1}\alpha(Z_{E1}T_x)\|_2^F \end{aligned} \tag{3.4}$$

There seems to be no over-fitting issue with the mathematical formulation given by Equation 3.4like the stacked encoder. The proposed HA approach is not a standard one and hence backpropagation won't work in this scenario. In order to find an efficient solution, we have used the ND-Adam optimization technique.

ND-Adam Optimization

This optimization technique helps in adjusting the input weights and parallely vanilla adam adjusts the other parameters. We divided the trainable parameters β into two parts namely: β^v & β^s such that $\beta^v = w_i$, $i \in N$ and $\beta^s = \beta \setminus \beta^s$ respectively. α^v_x and α^s_x denotes the learning rate for β^v and β^s. We update the value of β^v and β^s using the following algorithm [Zha18]:

Algorithm: ND-Adam

Initialization x <- 0

for $i \in N$:

$\quad w_{i,0}$ <- $w_{i,0}/\|w_{i,0}\|_2$

$\quad n_0(w_i)$ <- 0

$\quad v_0(w_i)$ <- 0

While $x < I$ (I denotes total iteration)

$\quad x$ <- $x + 1$

$\quad$ // Update β^v

$\quad$ for $i \in N$:

$\quad\quad \bar{f}_x(w_i)$ <- $\frac{\partial L}{\partial w_i}$

$\quad\quad f_x(w_i)$ <- $\bar{f}_x(w_i)$ - $(\bar{f}_x(w_i).w_{i,x-1})w_{i,x-1}$

$\quad\quad n_x(w_i)$ <- $\gamma_1 n_{x-1}(w_i) + (1-\gamma_1)f_x(w_i)$

$\quad\quad v_x(w_i)$ <- $\gamma_2 v_{x-1}(w_i) + (1-\gamma_2)\|f_x(w_i)\|_2^2$

$\quad\quad \hat{n_x}(w_i)$ <- $n_x w_i(1\text{-}\gamma_1^x)$

$\quad\quad \hat{v_x}(w_i)$ <- $v_x w_i(1\text{-}\gamma_2^x)$

$\quad\quad \bar{w}_{i,x}$ <- $w_{i,x-1}$ - $\frac{\alpha_x^v \hat{n_x}(w_i)}{\sqrt{\hat{v_x}(w_i)}+\epsilon}$

$\quad\quad w_{i,x}$ <- $\frac{\bar{w}_{i,x}}{\|\bar{w}_{i,x}\|_2}$

$\quad$ //Update β^s <- ADAM

$\quad \beta_x^s$ <- Update $(\beta_{x-1}^s;\beta_x^s,\gamma_1,\gamma_2)$

return β_I

The ND-Adam algorithm adjusts the learning rate to each input weight vectors and tries to maintain its magnitude as constant. It controls the learning rate precisely by optimizing only the direction. This tailored version overpasses the generalization gap present between the Adam optimizer and Stochastic Gradient Descent.

Content Based Filtering

We discussed about CBF in Chapter 1, there we learnt about the basic working principle of CBF. The content in CBF refers to the product/item specifications or attributes. In content-based filtering, the main objective is to include those key phrases to tag items, considering whatever the user likes, try looking up similar phrases in the repository and suggest different items with almost the same specifications. Greater the amount of information provided by user through implicit or explicit mechanism, stronger will be the user profile and correspondingly the power of recommendation will also increase. Figure 3.1 demonstrates a rough sketch of working of CBF in our proposed architecture. In our model, we recommended the set of items based on the attributes (movie genres) and the final order of recommendation was decided using 'release year' attribute of the movies in MovieLens dataset. In order to compute the similarity among different items, we used Pearson Correlation Coefficient (PCC) as discussed in the upcoming subsection.

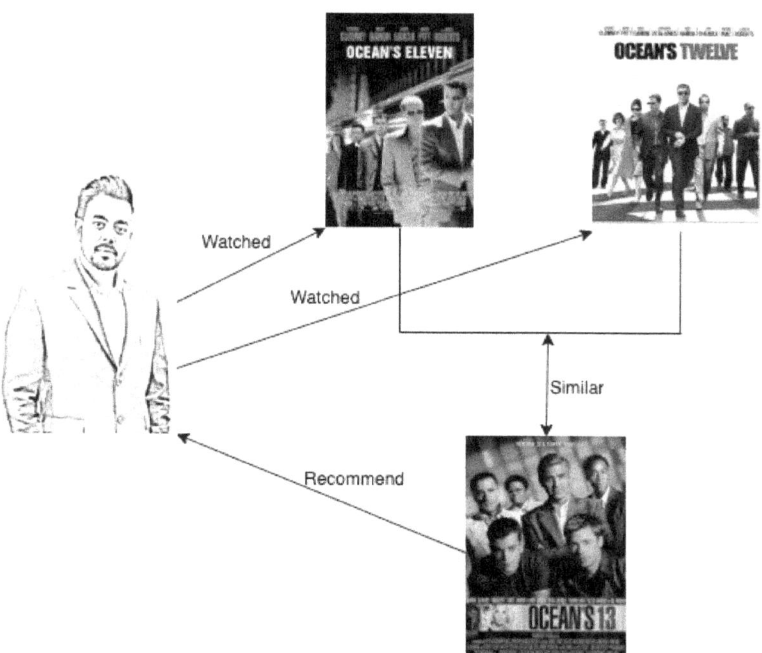

Figure 3.1 CBF based on attribute similarity

Pearson Correlation Coefficient (PCC)

Correlation indicates the relation between a set of data. One of the most famous approaches to measure correlation among the set of data is Pearson correlation. The correlation between items is computed in this measure on the basis of common item attributes. The value representation of PCC is given by:

+1: Total positive correlation
0: No correlation
-1: Negative Correlation
Mathematical representation of PCC is given by Equation 3.5.

$$PCC(A, B) = \frac{\sum_{i=1}^{n}(A_i - \bar{A})(B_i - \bar{B})}{\sqrt{\sum_{i=1}^{n}(A_i - \bar{A})^2}\sqrt{\sum_{i=1}^{n}(B_i -}} \qquad (3.5)$$

where, A and B represents two items.

Performance Measures

In this chapter, we have used several metrics to measure the performance of our proposed model. They are as follows:

1. Mean Absolute Error (MAE): It measures the absolute error present in the predictive model. It can be given by Equation 3.6.

$$\mathrm{MAE} = \frac{\sum |\mathrm{Predicted} - \mathrm{Actual}|}{\mathrm{Total\ Predictions}} \tag{3.6}$$

2. Root Mean Square Error (RMSE): It is a scale-dependent representation of square root of obtained error. It can be given by Equation 3.7.

$$\mathrm{RMSE} = \sqrt{\sum \frac{(\mathrm{Predicted} - \mathrm{Actual})^2}{\mathrm{Total\ Predictions}}} \tag{3.7}$$

3. Precision: The positive predictive value (Precision) measures, what fraction of positive assessments is correct? It is defined as Equation 3.8.

$$\mathrm{Precision} = \frac{\mathrm{True\ Positive}}{(\mathrm{True\ +\ False})\mathrm{Positive}} \tag{3.8}$$

4. Recall: It is also known as sensitivity. It measures, what fraction of actual positives were identified. It is defined as Equation 3.9.

$$\mathrm{Recall} = \frac{\mathrm{True\ Positive}}{(\mathrm{True\ Positive} + \mathrm{False\ Negative})} \tag{3.9}$$

Order of Prediction

A mesmerized order of recommendation degrades the satisfaction of users and the ultimate job of recommender system fails. Therefore, it is the utmost job of the recommendation model to recommend the items to the user in correct order. In this chapter, we have generated the final order of recommendation based on the release year.

For example: If the user 'x' has watched Ocean 11 and the system wants to recommend other ocean movies, then the correct order of recommendation for Ocean series will be:

Ocean 12 (Release Year: 2004) and then Ocean 13 (Release Year: 2007).

3.3 Proposed Architecture

In this section of the chapter, we highlighted the five different stages of our proposed architecture. The proposed model starts with data gathering and transformation and ends up by providing a suitable ordered list of items to be recommended to the end user. The complete flow diagram of the proposed architecture is illustrated by Figure 3.2. Detailed description of each stage is provided in the upcoming subsection of the chapter.

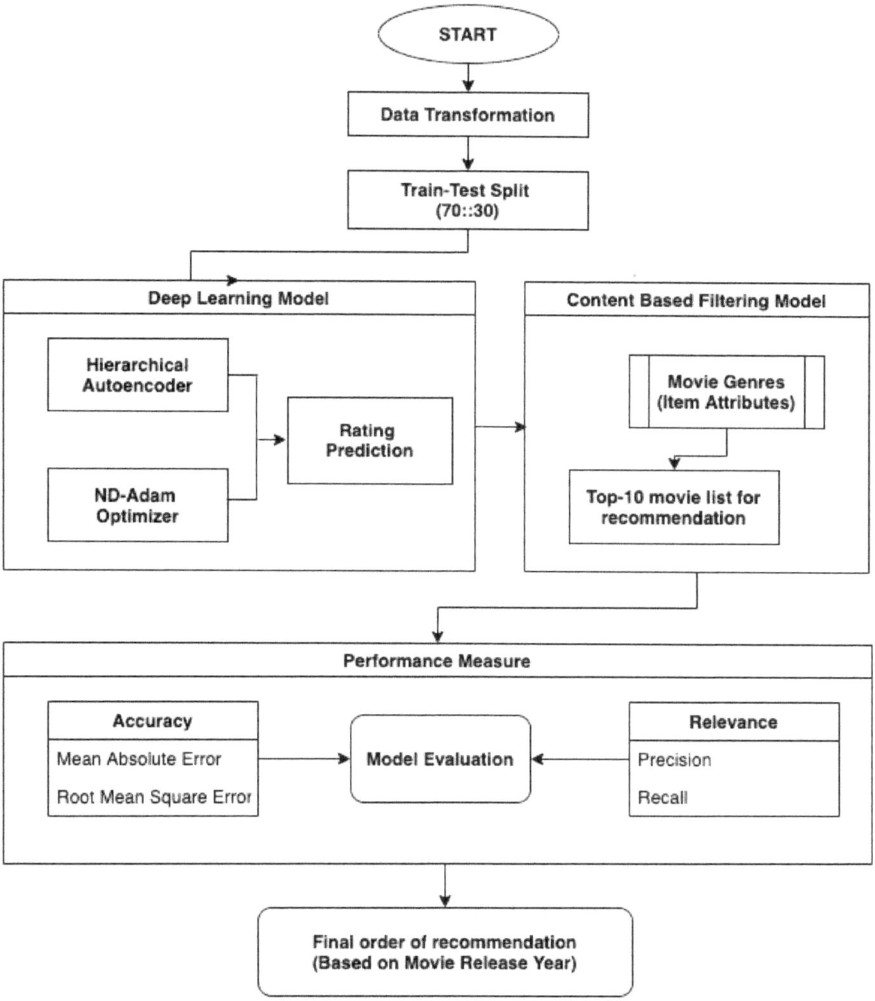

Figure 3.2 Flow model of proposed architecture

Data Collection, Transformation and Splitting

This step is a primary phase of any research field. It defines the procedure of collecting the data from different standardized sources in order to draw meaningful insights. In our proposed architecture, we have used different sizes of MovieLens dataset which is provided by the GroupLens research lab. The detailed description about the number of users, items, number of ratings, and other attributes are provided in Table 3.1.

We tested our model efficiency on all the datasets which are described in the table 3.1 in order to backup the claim of our model scalability. After successful collection of data, pre-processing is done. The transformed data has five columns namely: User_Id, Item_Id, Rating, Genres, and Release Year.

Table 3.1 MovieLens Dataset

Size	#Users	#Items	Scale	Sparsity	#Ratings
100K	943	1682	+1 to +5	93.7%	1,00,000
1M	6040	3900	+1 to +5	95.8%	10,00,209
10M	71,567	10,681	+1 to +5	98.7%	1,00,00,054
20M	1,38,493	27,278	+1 to +5	99.47%	2,00,00,263

User_Id: The unique ID of the user present in the system.
Item_ Id: The unique ID of the movie present in the system.
Rating: It denotes the rating score given to any item by the user. In case of MovieLens dataset, the scale of rating lies in the range of +1 to +5, where +1 reflects that the user is least satisfied with the item and +5 denotes that the user is highly satisfied with the item.
Genres: The genre information is represented in this column. In total, there are 19 genres present in the system. They are "Action, Adventure, Animation, Children's, Comedy, Crime, Documentary, Drama, Fantasy, Film-Noir, Horror, Musical, Mystery, Romance, Sci-Fi, Thriller, War, Western, and Unknown.
In our transformed dataset, we have represented this genre information in form of binary values, where 1 represents that the movie belong to that very genre and 0 indicates that the movie doesn't belong to that genre.
Release Year: It indicates the item/movie release year.
Once, the transformed dataset is ready, we split the dataset into training and test set. We use the training set to train the weights of our proposed model and then we use the test set to analyze the performance of the proposed model. The significance of training and test is illustrated via Figure 3.3.

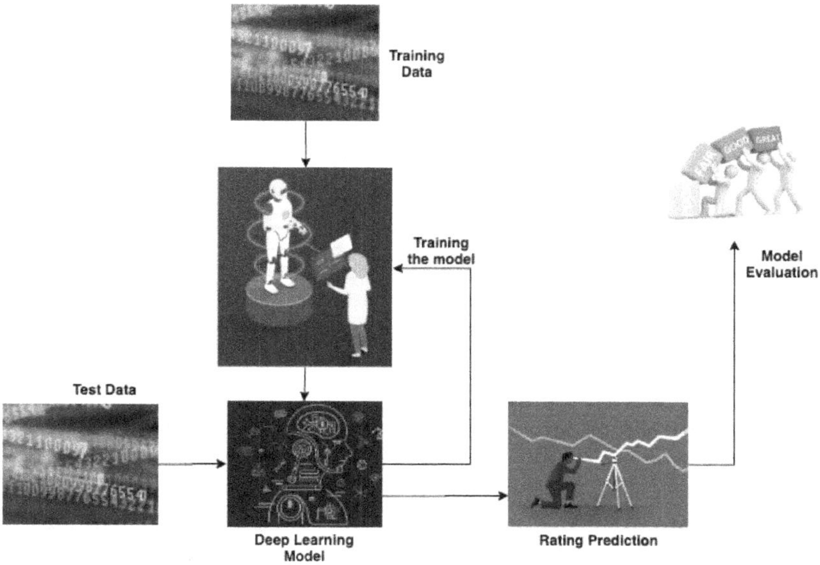

Figure 3.3 Training and Testing of the Model

Deep Learning Model

The hierarchical autoencoder has the power to learn decoder at every level. we used ND-Adam to optimize the weight values of the autoencoder. After achieving the optimal weight values, we used CBF to generate the Top-10 recommendations. In the learning phase of hierarchical autoencoder, we transformed the Equation 3.4 into lagrangian form by substituting:

$$Y_1 = \alpha(Z_{E1}T_X)$$
$$Y_2 = \alpha(Z_{E2}Y_1)$$
$$Y_3 = \alpha(Z_{E3}Y_2)$$
$$Y_4 = \alpha(Z_{E4}Y_3)$$

such that Equation 3.4 gets transformed into Equation 3.10.

$$\min_{Y_{E4},Y_{E3},Y_{E2},Y_{E1},Z_{D4},Z_{D3},Z_{D2},Z_{D1}} \|T_x - Z_{D4}Y_4\|_2^F + \|T_x - Z_{D3}Y_3\|_2^F + \|T_x - Z_{D2}Y_2\|_2^F + \|T_x - Z_{D1}Y_1\|_2^F \quad (3.10)$$

The Lagrangian imposes rigid equality throughout every iteration between both the parameters and its proxies. In reality, this isn't really essential. Only rigid equality across convergence can be required, or else the restriction can be laid back. In place of the Lagrangian, augmented approach provides a more functional solution. The partly sampled matrix 'M' scores are used for training purposes. The imputed ratings are obtained through each hierarchy until the weights of encoder and decoder are trained. The final assessment/rating is determined using a simple arithmetical mean by combining the outcomes of each hierarchy. The blueprint of hierarchical autoencoder is given by Figure 3.4.

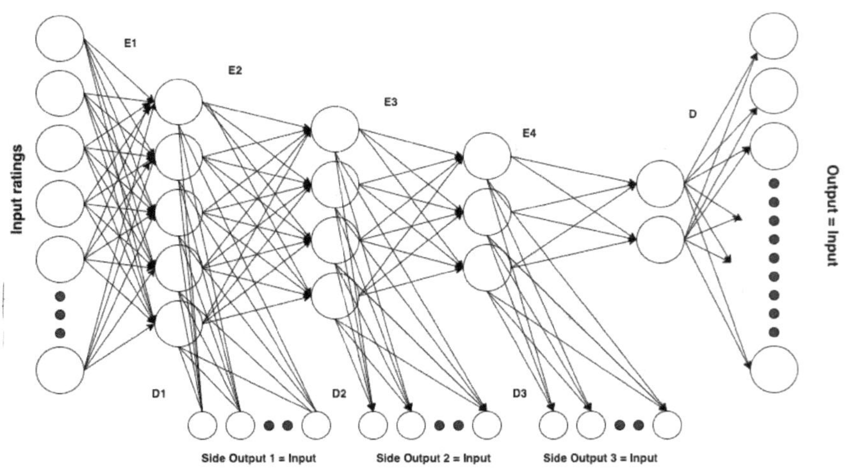

Figure 3.4 Hierarchical Autoencoder

Content Based Model

As discussed in section 3.2, CBF is responsible for generating the list of items to be recommended based on the user profile/ item attributes. Here in our proposed approach, we have considered movie categorical (genre) information to decide the Top-10 items to be recommended to the user.

For example: Suppose we have three item I_1, I_2, and I_3 which received high rating from the deep learning model. We want to recommend one item to the user 'X'. In such scenario, we will check what other items user 'X' has shown his interest. We compute the similarity of I_1, I_2, and I_3 with those set of items. The items which receive highest similarity in terms of sharing highly similar categorical information, that will be the one which will be recommended to user 'X'.

Similarly, in our model, we generate the Top-10 list of items for all the users in which he/she might be interested. The similarity between the items is calculated using PCC. Once we are ready with the recommendation list, we test the performance of our model.

Model Evaluation

In order to test the performance of our model, we evaluated the obtained results with respect to different quality measures such as MAE, and RMSE. We also measured the relevance of our model for Top-10 generated recommendation list using Precision, and Recall.

Final order of recommendation

This phase of the proposed architecture deals with providing the users with an ordered list of recommendation. Mesmerized order of recommendation generally degrades the user satisfaction. So, in our model we have arranged the generated recommendation list based on release year. The item release year plays a key role in making the recommender system aware about which item/movie was released after which item. If there is any series of movies, like Harry Potter and the Philosopher's Stone (2001), Harry Potter and the Chamber of Secrets (2002), Harry Potter and the Prisoner of Azkaban (2004), Harry Potter and the Goblet of Fire (2005), Harry Potter and the Order of the Phoenix (2007), Harry Potter and the Half-Blood Prince (2009), Harry Potter and the Deathly Hallows - Part 1 (2010), and Harry Potter and the Deathly Hallows - Part 2 (2011). In such scenario, the order of release matters alot. If the systems recommend 1st part of the movie, then 6th part, and then 2nd part, in such cases, the user will probably not return to the service provider and the service provider will loose a valuable customer just because of a small recommendation ordering mistake. The proposed model tried to take care of such issues and achieved a greater user satisfaction by providing an ordered list of items to be recommended to any particular user.

3.4 Experimental Outcomes and Observation

This section targets to represent the outcomes of each phase of proposed architecture. Before going into details about the experimental outcomes. We tried to present the exploratory data analysis of the MovieLens dataset via Figure 3.5 and Figure 3.6. In figure 3.5, we tried to analyze the user relation with the dataset. We made an analysis report of rating distribution with respect to the user. We also saw the different occupation to which the users might belong. Figure 3.6 displays the distribution of number of ratings and average rating. It provides an insight information about the movies and the different genre information.

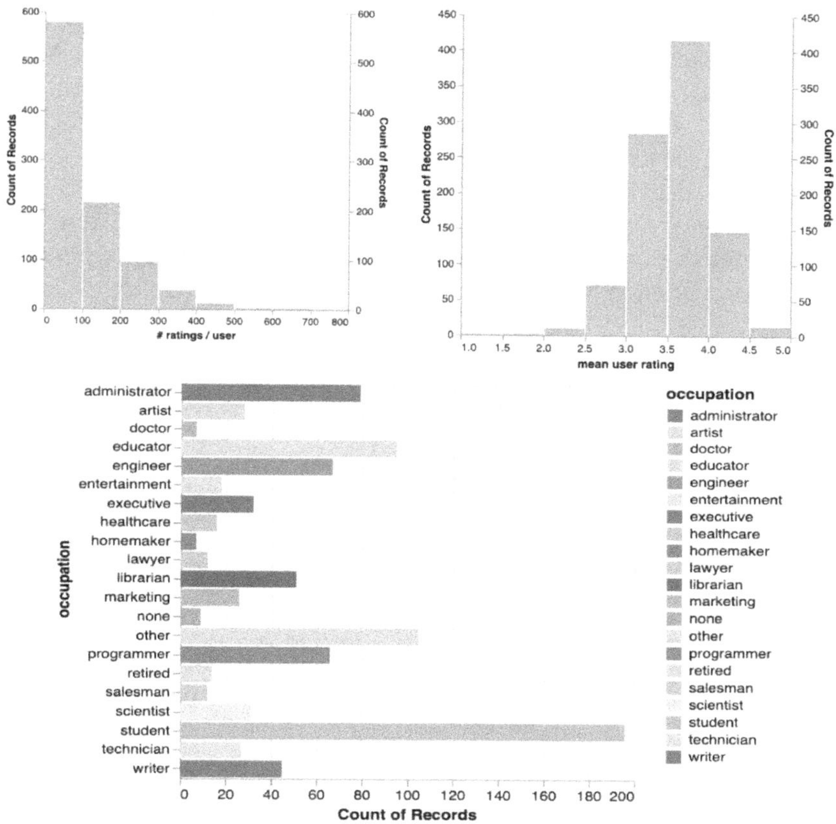

Figure 3.5 Rating distribution per User

We tested our model using four different size datasets namely: MovieLens 100K, MovieLens 1M, MovieLens 10M and MovieLens 20M. The detailed description of the dataset has already been discussed via Table 3.1. We measured the quality and relevance of the model using MAE, RMSE, Precision and Recall. The relevance of the model was measured after deciding the Top-N recommendation list generated by the content-based system. Higher value of precision and recall indicates better relevance. Similarly, low error value of MAE and RMSE indicates the better-quality measure achieved by the proposed model. The deep learning model being used in our proposed architecture need only single parameter to be specified. The value of encoder and decoder phases is effectively balanced. Even though there is no reason to support one another, we recommend that one parameter should be preserved and be equated to unity. We developed a four-layer architecture with 200-100-50-25 nodes. We carried out the experimentation with ASAE (Asymmetric Stacked Autoencoder), CDAE (Collaborative Denoising Autoencoder), MC (Matrix Completion), Probabilistic Matrix Factorization (PMF), and Stochastic Gradient Descent (SGD). The obtained experimental outcomes is demonstrated by Table 3.2, Table 3.3, Table 3.4, and Table 3.5. The best results are provided by our proposed architecture. It is assumed that the improvement is due to the fusion of supplementary information at each autoencoder level.

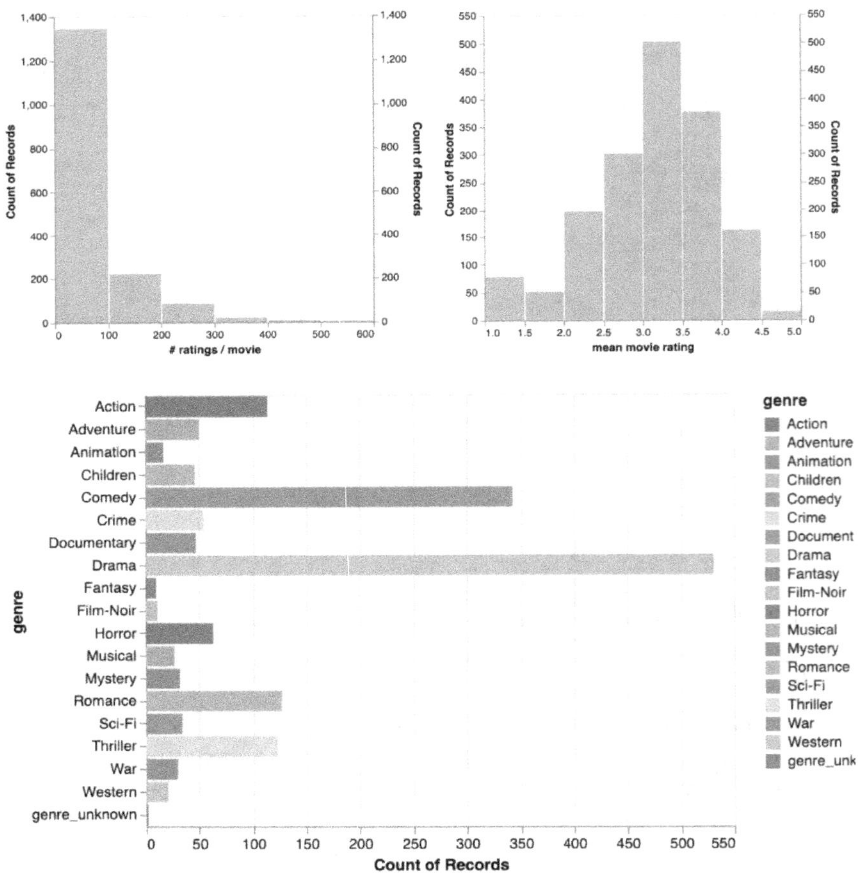

Figure 3.6 Rating distribution per Item

Table 3.2 Experimental Outcome for ML-100K

Approach	MAE	RMSE	Precision	Recall
ASAE [MT17]	0.742	0.953	0.510	0.642
CDAE[Wu+16c]	0.735	0.940	0.513	0.641
MC[GSM14]	0.732	0.938	0.515	0.641
PMF[MS07]	0.756	0.963	0.505	0.635
SGD[KBV09]	0.743	0.942	0.501	0.636
Proposed	0.7128	0.8821	0.526	0.644

Table 3.3 Experimental Outcome for ML-1M

Approach	MAE	RMSE	Precision	Recall
ASAE [MT17]	0.689	0.871	0.671	0.630
CDAE[Wu+16c]	0.691	0.878	0.671	0.625
MC[GSM14]	0.689	0.876	0.669	0.625
PMF[MS07]	0.724	0.912	0.630	0.603
SGD[KBV09]	0.743	0.942	0.632	0.606
Proposed	0.661	0.835	0.701	0.631

Table 3.4 Experimental Outcome for ML-10M

Approach	MAE	RMSE	Precision	Recall
ASAE [MT17]	0.673	0.847	0.502	0.615
CDAE[Wu+16c]	0.687	0.849	0.508	0.618
MC[GSM14]	0.684	0.852	0.511	0.614
PMF[MS07]	0.701	0.886	0.497	0.606
SGD[KBV09]	0.726	0.841	0.497	0.602
Proposed	0.634	0.832	0.503	0.615

Table 3.5 Experimental Outcome for ML-20M

Approach	MAE	RMSE	Precision	Recall
ASAE [MT17]	0.711	0.823	0.500	0.605
CDAE[Wu+16c]	0.703	0.811	0.513	0.601
MC[GSM14]	0.723	0.803	0.515	0.593
PMF[MS07]	0.716	0.821	0.511	0.598
SGD[KBV09]	0.721	0.838	0.495	0.600
Proposed	0.628	0.801	0.501	0.612

The precision and recall value show drastic change if we increase the value of 'N' in Top-N recommendations from 10 to 20 or 30. The obtained precision and recall reflects lower relevance with increase in 'N' value. Figure 3.7 and Figure 3.8 demonstrates the Precision and recall obtained for N = 10, N= 20 and N= 30. It can be clearly observed from the obtained Precision and Recall value that the proposed architecture attains high relevance when the N value in Top-N recommendation list is 10.

After generating the Top-10 items to be recommended, we decide the order of recommendation list based on the item release year. An instance of recommendation list generated for User-1 is discussed via Table 3.6. The correct order of recommendation for the generated recommendation list will be: Pinocchio(1940), To be or not to be(1942), It's a wonderful life(1946), Raising Arizona(1987), Dead Poets Society(1989), Homeward Bound: The Incredible Journey(1993), Two deaths(1995), Twelve Monkeys(1995), The Hunchback of Notre Dame(1996), and Mimic(1997). When the model encounters any two movies having a common release year, in such scenario, the movie that has high predicted rating will be recommended first.

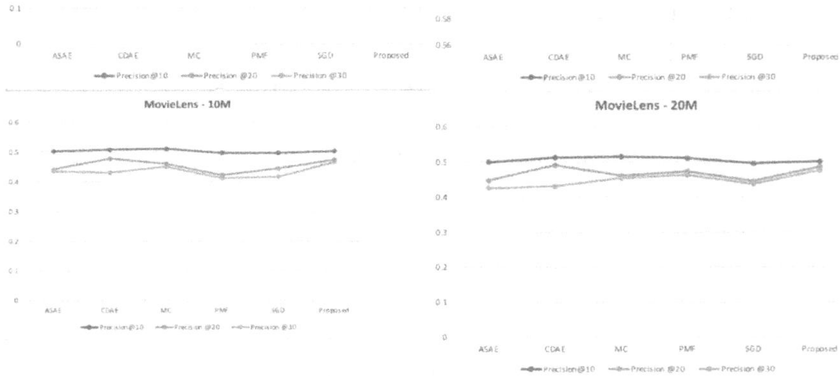

Figure 3.7 Precision @N = 10, 20, and 30

Table 3.6 Generated recommendation list for User-250 (ML-100K)

Movie Name	Release Year	Actual Rating	Predicted Rating
It's a wonderful life	1946	4	3.3
Mimic	1997	3	3.6
Homeward Bound: The Incredible Journey	1993	3	4
Twelve Monkeys	1995	4	3.6
Raising Arizona	1987	4	4.1
The Hunchback of Notre Dame	1996	5	4
To be or not to be	1942	2	3
Dead Poets Society	1989	4	3.4
Pinocchio	1940	4	2.8
Two deaths	1995	4	3.1

3.5 Results and Discussion

The proposed architecture yields best results in terms of quality and relevance. We also compared our proposed architecture with several other comparative models and it is found that the proposed model outperforms the existing content-based models. The proposed HA-CBRS model leveraged both the CB features and user-item rating relationship. It is believed that perhaps the improvement attained by the proposed architecture is due to the aggregation of supplementary information provided by each autoencoder level. The ND-Adam optimizer in our HA deep learning model helps in preserving the gradient direction. The comparative analysis report is discussed via Figure 3.9.

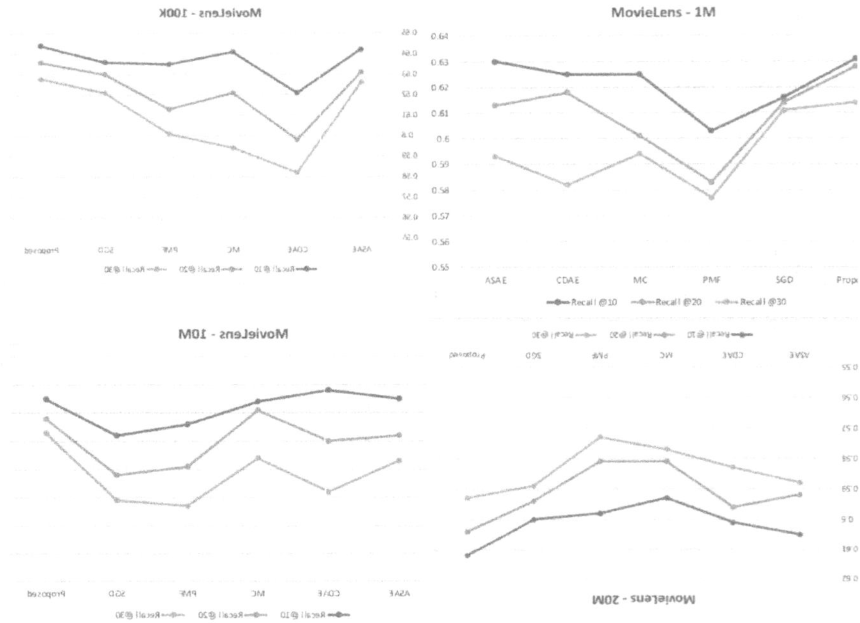

Figure 3.8 Recall @N = 10, 20, and 30

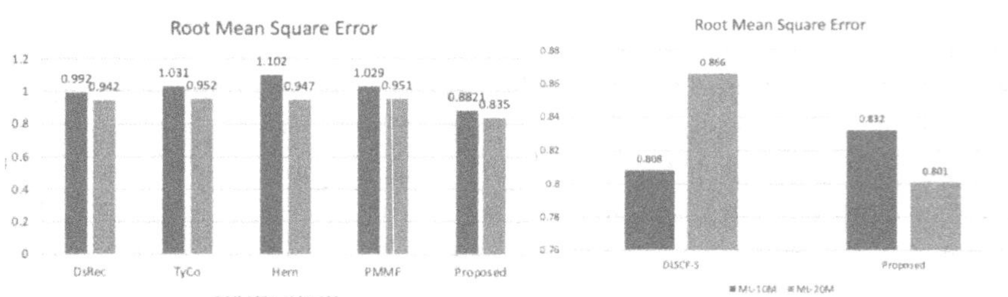

Figure 3.9 Comparative analysis of HA-CBRS

3.6 Conclusion & Future Direction

The main objective of this chapter was to build a recommender system that could generate an efficient recommendation list for different set of users with high accuracy (low error). In the past few years, autoencoders has displayed a massive improvement in prediction accuracy. Therefore, we have used hierarchical autoencoder as our base deep learning model in the proposed architecture. The obtained results on different sizes of MovieLens dataset shows the efficacy of our proposed model over several other state-of-art comparative models.

In future, we can make use of metadata information to target other factors to satisfy the users such as: Diversity, Novelty, and Serendipity.

CHAPTER 4

Recommendation with Deep Learning Techniques in Collaborative System

"Unity is strength. When there is teamwork and collaboration, wonderful things can be achieved" — Mattie Stepanek

Models focused on collaborative filtering (CF) should explain the associations or similarities between users and items undergoing analysis. In order to encourage recommendations, CF uses concurrent similarities among both users and items. This makes it possible to make serendipitous recommendations. In other words, CF models will strongly suggest any item 'I' to user 'X' based on interests of a highly similar user 'Y'. In addition, embedding can be instantaneously learned without any need for hand-crafted features. In this chapter, we constructed a deep learning collaborative model by blending the explainable RBM in deep belief networks backed with a new similarity measure of divergence. The parameters of the new similarity measure are optimized using a nature inspired Marine Predator algorithm. The proposed model is tested on 6 benchmark datasets (BookCrossing, Epinions, Flixster, FilmTrust, MovieLens-100K, and MovieLens-1M) to demonstrate the efficacy of the model over state-of-art comparative models.

4.1 Introduction

Recommender system is a screening mechanism for information that is widely used to help the end customer navigate through a significant pool of items which are only of user interest. It collects several facts about the customer's behaviours, inclinations, preferences or tastes, and uses this data obtained with the aid of feedback to suggest item to potential customers. In our daily lives, we usually call upon people with identical thoughts, tastes and interest or other trustworthy outlets for their opinions on the existence and condition of assets to receive proposals for a commodity or material. The online e-commerce sites like AMAZON uses RS to facilitate this word-of-mouth concept, and sites such as Netflix uses it to suggest movies to customers by offering personalized recommendations.

The most popular and widely-used solution in RS is CF-based system. They rely on computing similarities between the users or items present in the system. The working principle of CF is that if a person seems to have the similar preference for a certain amount of products in the past, perhaps in the future they would have a similar taste. The preference for such commodities is much more nuanced and impossible to classify. That is because the user interests often cannot be described by merely using keywords or by consistency. The preference, action or taste of other similar users may be used to obtain the interests of target user. In order to compute the similarity between the users or items, several researchers have suggested the use of different similarity metrics [SD19a] such as Cosine similarity, PCC (Pearson Correlation Coefficient), Jaccard similarity, singularity-based similarities [BOH12], significance-based similarities [Bob+12], etc. Sparsity, however, is one of the key failings of this popular strategy. This challenge arises implicitly in the system and is due to a rising variety of consumers and products. Owing to this, many users can only review or purchase a few items from the total items available. In fact, very few consumers may have purchased or judged even highly well recognised products. Similarities among users that lead to sparse utility (user-item rating) matrix are therefore difficult to compute. This has a great impact on the reliability of designed recommender system. However, there is a potential loss of sheer determination of similarities because of the missing knowledge processed by the traditional method. In this chapter, our one of the objectives is to address this issue by using a new similarity metric based on mean measure of divergence & co-rated items for finding the similarity between the users present in the system.

There can be some advantages to fundamental explanations for specific recommendations, such as: simply helping the consumer make the right choice (precision), helping the consumer make a quicker decision (efficacy), and disclosing the logic behind the suggestion of the system (transparency). As an outcome, people are more inclined to obey the advice and utilise the method in better ways. For example, by listing related movies received from the user's social media platform, the Netflix recommender framework justifies its movie recommendations. The recommendation framework from Amazon displays identical items to those ordered or seen by the consumer (and perhaps other similar consumers) while suggesting a new item utilizing CF system. The Restricted Boltzmann Machines (RBM) approach can allocate a low-dimensional variety of features to products in a latent space. The new set of features contain the user's preferences and aspects that they admire, but these are complicated to understand. Therefore, in such models, the basis for the suggestion or the logic behind the suggested product is not evident. Recently, RBM methods have proven to be effective in developing deep learning strategies in diverse datasets to learn and forecast trends since they can produce very detailed outcomes. They do, however, suffer from inadequate understanding of the findings, especially in the case of recommendation systems. Lack of explanations will contribute to users not trusting the recommendations made by the system. Therefore, by accepting it is the only way for the consumer to judge the reliability of a recommendation. This, though, is contradictory to one of the purposes of a recommendation framework, which decreases the time spent on exploring products by consumers. Designing advice mechanisms that will include reliable advice that, at the same time, promote the conveyance of the logic behind the recommendations to the consumer will be both useful and helpful. Choosing an explainable procedure with reasonable predictive performance or a more precise technique, such as RBM, that does not provide explanatory recommendations, is a major challenge in designing a recommendation method. In this chapter, we have addressed this issue by introducing explainable component in the deep belief network. Detailed explanation of the deep belief network used in our proposed framework is discussed in the upcoming section of this chapter.

Following are the significant contributions supported by our work in this chapter:

1. Usage of new similarity metric based on mean measure of divergence for capturing the rating behaviour of users efficiently in a sparse environment.
2. Optimizing the parameters of the new similarity metric using a metaheuristic approach namely: Marine Predator Algorithm.
3. Generating an explainable set of recommendations using Deep Belief Networks (DBN) that comprise multiple layers of RBM.

4. Testing the proposed framework efficacy on 6 datasets using Root Mean Square Error (RMSE) and Mean Absolute Error (MAE).
5. Recommending the top-10 list of items to the test users based on the rating values obtained for the unrated set of items present in test set.

The remaining section of the chapter is organized as: Section 2 highlights the elementary concepts and principle techniques that worked together to build the proposed architecture. The flow diagram of proposed framework is discussed in Section 3 of the chapter. Section 4 talks about the outcomes obtained at each stage of the proposed architecture. The obtained results and comparative analysis are performed in Section 5. This chapter also provides future direction for the implemented approach along with the conclusion in Section 6.

4.2 Experimental Methodology

To solve the inherent shortcomings of the CF-based models and build a more robust learning process, two key components that need to be tackled are: i.) Inefficient correlation among the users in case of sparse datasets, and ii.) Design a robust deep neural network architecture for CF-based recommender engines that can recommend explainable items. In our proposed model, we have blended different methodologies to tackle the aforementioned issues.

Collaborative Filtering

To overcome the shortcomings of CBF, collaborative filtering integrates similarities between consumers and products simultaneously. CF-based models allow for wondrous recommendations for related users, that is, recommendations based on a resemblance between two users. Besides this, the embeddings can be taught automatically, something that cannot be achieved by hand engineering. Let us consider the scenario of movie recommendation where the feedback about any movie present in the system lies in either explicit or implicit category.

1. Implicit: If any customer watches any movie, then the system assumes that the customer is actively interested in that movie.
2. Explicit: Customers indicate their interest in any movie by providing a rating score.

To simplify, let all variables would be binary representing whether the person is interested in the movie or not. When a user hits the home screen, the framework can suggest movies that are important to the user's interests based on two factors:

1. Similar movies in which the user has showed interest in the past.
2. The set of movies in which the similar user has shown interest.

The following illustrations shows the 1-dimensional and 2-dimensional embedding of the consumer and products. Assigning to each movie/product a scalar value in range of [-1, 1] expresses whether certain content is suitable for children [-1] or adults [+1]. Suppose we also allocate scalar values to each consumer in range of [-1, 1] that represent their degree of interest in an adult or children's movies. The result of the product embedding and the consumer embedding varies based on the kind of content a consumer would enjoy. Figure 4.1 and Figure 4.2 demonstrates the 1-D embedding and 2-D embedding of the aforementioned scenario respectively. One attribute of the product wouldn't be enough to clarify everybody's tastes. To address this issue, we need to understand the indicator of a product being a blockbuster or perhaps an arthouse film. With a second attribute, the framework is now able to integrate the 2-D embedding for each movie.

Figure 4.1 1-D Embedding of Users and Movies

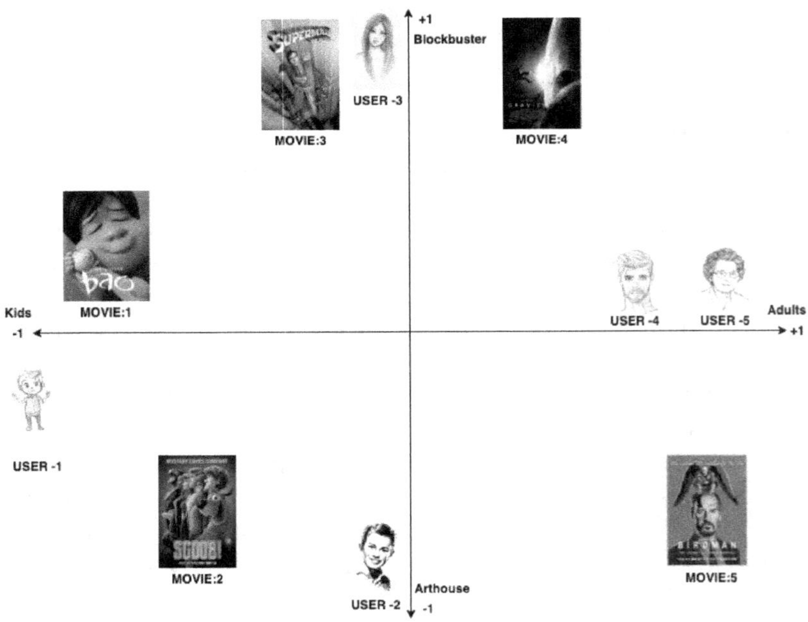

Figure 4.2 2-D Embedding of Users and Movies

The collective essence of this strategy is seen as the model understands the embeddings. Suppose fixed embedding vectors are used for the movies. The model will then learn the embedding vector for users' interests to better describe the choices they made. The shared users with identical interests are grouped together. Likewise, if we gain knowledge embeddings that best describe the feedback matrix, then we will use the term embeddings to help identify movies. There would be strong embeddings since products that are enjoyed by related users are relatively close throughout the embedding area.

Outline of new similarity metric

The novel similarity metric is intended to increase the consistency of the existing recommendation system in comparison to their efficiency. To show the reliability and simplicity of the metric, a series of trials was performed on different datasets of varying sizes. The proposed similarity measure is based on two principle components: *i.)* Mean measure of divergence, and *ii.)* Similarity based on co-rated items and weight. The traditional resemblance test doesn't fit well in sparse environments since it doesn't use relative scores efficiently. This motivated us to propose a new similarity measure that can deal in such scenarios. Each consumer possesses their own behaviour for voicing their preferences. A certain consumers seem to score more positively or negatively than the others. User biasness of scores interferes with the users' preference levels. The conventional method neglects this aspect. Means of calculating divergence is used to build a likeness index in which users' scores are contrasted to one another. The mean measure of divergence between a pair of users can be computed by using Equation 4.1.

$$\mathrm{Sim}(X,Y)^{MMD} = \frac{1}{1 + \left(\frac{1}{r}\sum_{i=1}^{r}(\theta_X - \theta_Y)^2 - \frac{1}{|I_X|} - \frac{1}{|I_Y|}\right)} \quad (4.1)$$

where, I_X and I_Y represents the total number items rated by user 'X' and user 'Y' respectively. θ_X and θ_Y denotes the user 'X' and user 'Y' rating vector. The total number of co-rated items by user 'X' and user 'Y' is denoted by 'r'.

For example: Suppose the rating scale of the system lies in the range of [1,5], such that min=1 (least interested), and max=5 (most interested). Let I_X and I_Y be denoted by:

I_X = (2, 4, 5, 0, 1, 3, 0, 4, 4, 1)
I_Y = (1, 5, 4, 2, 4, 0, 4, 3, 2, 5)
min = 1, and max = 5
$\theta_X = (\theta_{\#min}, \ldots, \theta_{\#max})$

$\theta_X = (\theta^1, \theta^2, \theta^3, \theta^4, \theta^5)$
$\theta_X = (2, 1, 1, 3, 1)$
$\theta_Y = (\theta^1, \theta^2, \theta^3, \theta^4, \theta^5)$
$\theta_Y = (1, 2, 1, 3, 2)$

By putting the value of I_X, I_Y, θ_X, and θ_Y, we can compute the similarity between user 'X' and user 'Y' based on mean measure of divergence. To enhance the essence of the computed similarity metric, we incorporated weight and value parameter to the mean measure of divergence. Equation 4.2 demonstrates the blueprint of new similarity metric.

$$Sim(X,Y)^{new} = Sim(X,Y)^{MMD} + \frac{1}{max - min + 1} \sum_{i=0}^{max-min} W^i.V^i_{X,Y} \quad (4.2)$$

where, V and W denotes the value and weight respectively. The weight 'W' lies in the range of [-1,1] and V represents the set of commonly rated items.

For example: $I_X = (2, 4, 5, 0, 1, 3, 0, 4, 4, 1)$
$I_Y = (2, 5, 4, 2, 4, 0, 4, 3, 2, 5)$
min = 1, and max = 5
$V_{X,Y} = (V^0_{X,Y,}, .. V^{max-min}_{X,Y})$
$V^i_{X,Y} = \frac{a}{b}$, where a = absolute difference in the rating scores given by user 'X' and user 'Y', and b = Total number of commonly rated items by user 'X' and user 'Y'.
$V_{X,Y} = (1/10, 3/10, 2/10, 2/10, 2/10,)$ and $W^i = (W^0, W^1, W^2, W^3, W^4)$
$W^i = ([1, 0.6], [0.6, 0.2], [0.2, -0.2], [-0.2, -0.6], [-0.6, -1.0])$

Using the calculated values of weight and value parameter, we compute the similarity between user 'X' and user 'Y' with better significance. The optimal values of the weight parameter of Equation 4.2 are computed using one metaheuristic approach namely: Marine Predator Algorithm. Detailed description of Marine Predator Algorithm is discussed in the upcoming subsection of the chapter.

Marine Predator Algorithm (MPA)

It is a nature-inspired optimization approach that implements the standards that generally regulate optimum foraging strategy as well as encounter rates shared with predator and prey in marine environments. Generally, the foraging behaviour of several species in the real world is a strictly random mechanism; a probabilistic model wherein the next movement is based on the current location, as well as the transition likelihood to the next position can be modeled mathematically. Studies have shown that many species use the Lévy flight characteristics, which is their preferred foraging strategy. Lévy flight could be the most effective search strategy for prey in low population densities in natural habitats. According to research [Far+20], the best approach for exploring and foraging is to consider Lévy and Brownian combination. When hunting for prey in a prey-sparse region, Lévy strategy is prevalent. However, when hunting for prey in a prey-abundant region, the strategy is widely changed to Brownian approach. The study concludes that a marine predator needs to spend the bulk of its time in seafloor environments exhibiting half of time using Brownian motion & other half using Lévy motion during their life. There has been some research that showed that marine predators thrive via cognitive skills assisting in their hunting, mating, and selection of habitat. This results in the following general principles for optimum foraging and cooperation and memory in marine predators:

1. In case of dense and sparsely populated prey environment, marine predators employ Levy and Brownian movement respectively.
2. They demonstrate the same levels of Levy and Brownian movement during their lifespan switching between diverse habitats.
3. Due to the normal and human-caused environmental impact, they modify behaviours to locate places of prey distribution that is distinct from others.
4. For velocity = 0.1 :: Ideal hunting tactic is Levy movement, prey movement can be either Levy or Brownian.
5. For velocity = 1 :: Ideal hunting tactic is Brownian movement, prey movement can be Levy.
6. For velocity (high; >=10) :: Ideal hunting tactic is to not move at all, prey movement can be either Levy or Brownian.

7. They have excellent memories of remembering their colleagues as well as how to get to food sources.

In the Brownian motion the phase(step) length is extracted from a Gaussian distribution with $\mu = 0$ and $\sigma^2 = 1$, where μ and σ^2 denotes mean and variance respectively. The density function at point 'a' for Brownian motion can be computed using Equation 4.3.

$$F_B(a; \mu, \sigma) = \frac{1}{\sqrt{2\pi}} \exp(-\frac{a^2}{2}) \tag{4.3}$$

The density function at point 'a' for Levy motion can be computed using Equation 4.4.

$$F_L(a, \alpha, \gamma) = \frac{1}{\pi} \int_0^{\infty} \exp(\gamma q^{\alpha}) \cos(qa) dq \tag{4.4}$$

where, α: distribution index, γ: for selecting scale unit.
$\alpha = 1$: Cauchy Distribution
$\alpha = 2$: Gaussian Distribution

The following pseudo-code demonstrates the working of MPA:

MPA: Marine Predator Algorithm

Notation

C_i: Current Iteration
M_i: Maximum Iteration
$\vec{S_i}$: Step size
$\vec{E_i}$: Elite Index
$\vec{R_B}$: Random number based on Brownian
$\vec{R_L}$: Random number based on Levy
$\vec{P_i}$: Prey ; i=1..n
n: Total population
$\vec{B}$: Binary vector
$\otimes$: Entry-wise multiplication
C: 0.5 (Constant)
$\vec{R}$: Random number $\in[0,1]$

Initialization

Initialize P_i

While(!TerminationCriteria)

Fitness calculation
Elite matrix construction
Memory Saving
If($C_i < \frac{M_i}{3}$)
$\vec{S_i} = \vec{R_B} \otimes (\vec{E_i} - \vec{R_B} \otimes \vec{P_i})$
$\vec{P_i} = \vec{P_i} + C.\vec{R} \otimes \vec{S_i}$

Else If($\frac{M_i}{3} < C_i < 2 * \frac{M_i}{3}$)
$\forall$ i $\in$ [1..n/2]
$\vec{R_L} \otimes (\vec{E_i} - \vec{R_L} \otimes \vec{P_i})$
$\vec{P_i} = \vec{P_i} + P.\vec{R} \otimes \vec{S_i}$

$\forall$ i $\in$ [n/2..n]
$\vec{S_i} = \vec{R_B} \otimes (\vec{R_B} \otimes \vec{E_i} - \vec{P_i})$
$\vec{P_i} = \vec{E_i} + P.CF \otimes \vec{S_i}$
$CF = (1 - \frac{C_i}{M_i}^{2\frac{C_i}{M_i}})$

Else If($C_i > 2*\frac{M_i}{3}$)
While($C_i > \frac{2}{3}M_i$)
$\vec{S_i} = \vec{R_L} \otimes (\vec{R_L} \otimes \vec{E_i} - \vec{P_i})$
$\vec{P_i} = \vec{P_i} + P.CF \otimes \vec{S_i}$

End (If)
Update Elite
Save Memory
$\vec{P_i} = \vec{P_i} + CF\ [\vec{X_{min}} + \vec{R} \otimes (\vec{X_{max}} - \vec{X_{min}})] \otimes \vec{B}$; r<=FAD
$\vec{P_i} = \vec{P_i} + [FAD\ (1-r) + r](\vec{P_{r1}} - \vec{P_{r2}})$; r>FAD

END

RBM with Explainability

Explanability in recommender system provides a threefold benefits: *a.)* It ensures efficiency, *b.)* It enhances effectiveness, and *c.)* It maintains transparency. Consequently, the decision is more likely to be followed by consumers, which makes for a smoother user experience. Model-based systems such as RBM tends to be black-box system. RBM approaches can produce powerful outcomes for the ability to learn and identify trends in high dimensional datasets. As it comes to recommender schemes, though, they suffer from the lack of interpretation of the outcome. Not providing adequate descriptions will lead to consumer distrust of recommender system specific suggestions. It will be quite advantageous to design recommender programs that have reliable advice, and enable the customer to understand their justifications at the very same time as they end up making them. In this chapter, we used an additional component in traditional RBM to justify the explainability in the list of generated recommendation list.

Generally, the explainable mechanism can be divided into three categories namely: Influence-Style (Item-based), User-Style, and Keyword-Style. We have used influence-style that aims at presenting the items that contributed most on RS's decision. The user-style and keyword-style are based on based on similar users and demographic data / item attributes respectively. Performing the recommendation activity using explainable objects could yield better interpretable results, and somehow satisfied with the power of model-based CF that would enhance prediction power.

The RBM architecture comprise of two layers: Visible and Hidden Layer. Both layers of the stochastic NN interact with each other. There is no communication between the neurons present in the same layer. The neurons at the lower layers transfer information up to the upper layers, so the upper layers return information to the lower layers. RBM makes multiple passes between both the visible and hidden layers, producing a generative probabilistic model such that reconstructions are the same as the initial input. More precisely, the RBM seeks to make a generative model which allows them to predict whether a movie would be liked by a user who has never seen it before by taking into account how close the user is to those who have seen it and those who have given it a rating. The number of neurons at visible layer will be equal to 'M' (Total number of movies present in the MovieLens dataset). The rating score at each layer will have a normalized value between 0 and 1, where '0' represents that the user has not seen the movie and '1' denotes that the user is highly interested in the movie. Neurons in the upper layer will interact with neurons in the next layer to understand the implicit desires about users-movies interaction. A probabilistic value is computed and assigned to each pair of visible and hidden vector. The probability value can be calculated using Equation 4.5.

$$P(V, H) = \frac{e^{-E(V,H)}}{N} \tag{4.5}$$

where, E: System energy, and N: Normalizing factor. An explainable component is added to the aforementioned RBM architecture. The illustration of RBM with explainability is demonstrated via Figure 4.3. If the question has a way of being understood, then it can be explained; if it has a way of being answered, it can be answered. The concept is that because many other users have given the suggested item a rating, this will give a baseline for a recommendation, utilizing user-style techniques. Explanability score can be computed using Equation 4.6, where 'x' denotes an item and 'u' represents a user.

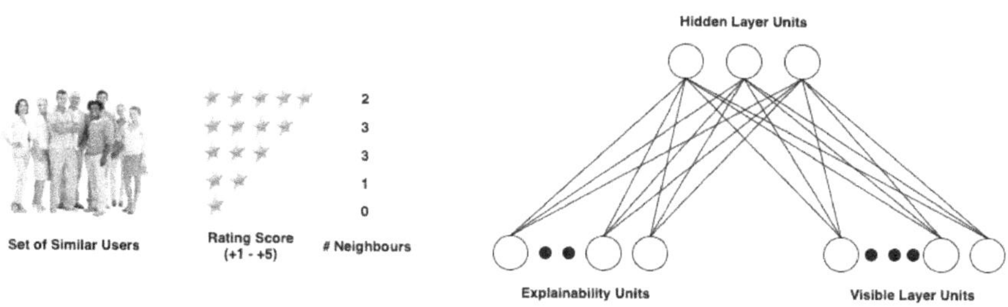

Figure 4.3 RBM with explainability

$$E(u, i) = \frac{\sum_{i \in N(u)} R_{i,x}}{\sum_{N(u)} R_m} \tag{4.6}$$

where, N(u): Set of similar users with respect to user 'u', $R_{i,x}$: Rating received by item 'x' from user 'i', and Rm holds the maximum rating provided by the similar set of users N(u) to item 'x'. In this chapter, we decided the neighbours by using a new similarity metric which is based on mean measure of divergence and weight-value parameter for capturing user behaviour. If the user has not rated the particular item 'x', we consider the value as '0'(missing rating score) to represent it. The value obtained from Equation 4.6 lies in the range of [0, 1], such that 0 denotes no explainability and value > 0 represents that item 'x' is explainable for user 'u'.

Joint distribution over visible and hidden layer can be defined with explainability scores (e) by using Equation 4.7 [Hin12].

$$
\begin{aligned}
P(H_j = 1|V, e) &= \sigma(b_j^1 + \sum_{i=1}^{n} V_i W_{ij}^1 + \sum_{i=1}^{n} e_i W_{ij}^2) \\
P(V_i = 1|H) &= \sigma(b_i^2 + \sum_{j=1}^{m} H_j W_{ij}^1) \\
P(e_i = 1|H) &= \sigma(b_i^3 + \sum_{j=1}^{m} H_j W_{ij}^2)
\end{aligned}
\tag{4.7}
$$

where, b^1, b^2, b^3: Bias
m, n: #Hidden & #Visible units
W^1, W^2: Entire network weights
$\sigma(X)$: logistic function

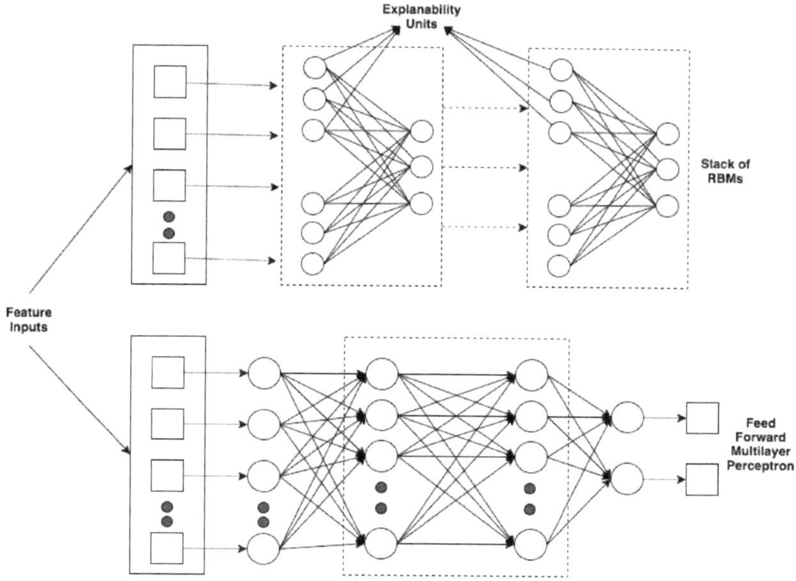

Figure 4.4 Architecture of Deep Belief Network

Deep Belief Network (DBN)

A DBN has several RBMs stacked together to capture the complex representation of training data. The information is incorporated into several layers in deep belief networks, where each of those layers knows the whole set of input data. Feature extraction is done in the first layers of CNN, while more layers use the fundamental characteristics, such as edges, to extract several patterns, such as complex features. DBN at the other side, control the layers, while still being robust and invariant at each time. In DBN, two major roles played by the nodes present at hidden layer includes: a.) Act as visible to the nodes that proceeds the layer, and b.) To the preceding nodes, act as hidden. These entities help the system in finding patterns within the data. It is standard procedure to train DBN using greedy learning approach. This approach attempts to optimize and fine-tune the generative weights by traversing from the lowest layer and moving up to higher layers. On a layer-by-by-layer basis, which means one layer of the DBN is fully trained at a time. Thus, each level is receiving a different data, and is using the obtained results from the last layer as its input. Training DBN using greedy algorithms can be fast and effective. Moreover, they are helpful in strengthening the adjustment of each layer weights. The working of DBN can be categorised into two phases namely: Pre-train and Fine-tune phase. The Pre-train phase comprise of stack of RBMs as discussed in previous subsection. We have incorporated explainability component with the stacked RBM. The Fine-tune phase comprise of a simple feed forward network. Figure 4.4 illustrates the architecture of DBN used in this chapter.

The original data is first ingested into the lower layer before the actual learning is performed by using an RBM. After this, the parameters are set, the RBM is applied to the hidden nodes of the model in the second layer. When the procedure is over, a fitting model is found at the upper-subnet layer, features are extracted. Also, additional networks of supervised learning may be added after the unsupervised learning task to implement supervised tasks such as classification.

Performance Measures

In order to test the performance of the proposed model, we have used two metric namely: Mean Absolute Error (MAE), and Root Mean Square Error (RMSE). The description of these techniques has already been discussed in the previous chapters. Both the techniques are statistical approach for measuring the deviation between actual and predicted rating.

4.3 The Proposed Architecture

In this section of the chapter, we have discussed about the several phases of our proposed recommendation framework. The model starts with pre-processing of data and ends with a final list of recommended items to any particular user 'u'. Figure 4.5 demonstrates the workflow of the proposed architecture. In the first phase of the proposed architecture, we transform the MovieLens dataset in such a way that the rows represent the items and column represents the users present in the system. We perform min-max normalization on the rating scores, in order to pull down the rating behaviour of users on a common scale ranging from 0 to 1. Generally, we can categorise the set of users present in the system into two categories namely: Tough and Easy raters. Tough raters are those who are very strict in giving ratings and they barely give high rating to any movie, unless they are fully satisfied with the movie. Whereas, easy raters are those users who are very lenient in giving ratings to any movie. Min-max normalization plays a key role in normalizing the behaviour of all such users present in the system. The min-max normalization can be computed by using Equation 4.8.

$$R' = \frac{R - min_u}{max_u - min_u}(max_s - min_s) + min_s \tag{4.8}$$

where, R': New rating value
R: Current rating score given by user 'u' to item 'i'
max_u: maximum rating given by user 'u'
min_u: minimum rating given by user 'u'
max_s: 1 [upper limit of normalized rating score]
min_s: 0 [lower limit of normalized rating score]

After pre-processing the data, we split the dataset into two divisions: Training set (70%) and Test set (30%). The training set is used to train the proposed model and then we check the performance of our model using the test set.

The second phase of the proposed model deals with calculation of similarity between the items by using a combination of MMD and (weight, values) parameters. This new similarity computation technique used in this chapter proves to be efficient even in case of sparse environment. We decide the optimal weights of the new similarity technique using a metaheuristic approach namely: Marine Predator Algorithm (MPA) in the third phase of the model. MPA makes use of both Levy and Brownian motion to converge towards the optimal solution.

Once we are ready with similarity index and item-based neighbours, in the fourth phase, we compute the ratings of unrated items using DBN which further comprise of explainable RBMs and feed forward multilayer perceptron. The explainability component helps in backing up the reasons behind recommending a particular item to any individual.

The performance of the model is computed by using statistical methods like RMSE and MAE which measures the deviation of actual rating scores from the predicted scores. It measures the error value in the predictions. The lesser the RMSE and MAE value, better is the model performance.

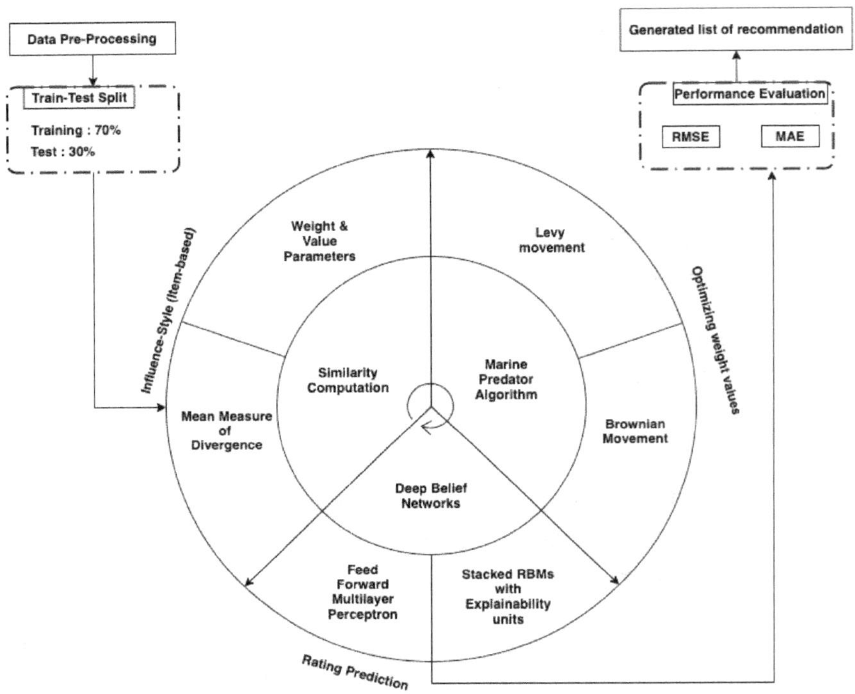

Figure 4.5 Workflow of proposed CF-DBN architecture

4.4 Experimental Outcomes and Investigation

This section of the chapter deals with a thorough investigation of outcomes obtained at each phase of the proposed model. This chapter has tested the proposed architecture on 6 different datasets having different set of users and diverse item lists. The different sparsity percentage of each dataset helps in justifying the efficiency of new similarity metric based on MMD and other parameters such as weight & values.

1. Bookcrossing: This dataset comprises of approx. 1.1M preference records of around 100K users. Total number of items present in the system is around 340K.

2. Epinions: This is a real-world dataset that has around 71K users & 104K items. Total number of rating scores present in the dataset is around 575K.
3. Flixster : It is a dataset which is widely used for testing any recommender framework. It consists of 1M users and 49K items. Users with ratings is around 150K and the total number of rating scores present in the dataset is around 8.2M.
4. Filmtrust: Users: 1592, Movies: 1930, Ratings: 28645, Scale: 1-10, and Sparsity: 99.06%.
5. MovieLens-100K : Users: 943, Movies: 1682, Ratings: 100K, and Sparsity: 93.7%.
6. MovieLens-1M : Users: 6040, Movies: 3900, Ratings: 1M, and Sparsity: 95.8%.

We performed the pre-processing of the data before dividing it into training and test set. We transformed the dataset into appropriate form. We also normalized the rating scores. The output of transformation and description of rating scores is illustrated by Figure 4.6. After transforming the dataset, we normalized the book_rating by using Min-max normalization. Similarly, we transformed all the other 5 datasets and normalized it such that they lie in the range of [0,1]. Then we splitted the dataset into training and test set. The train set is used to further compute the similarity between the items and at the same time we are deciding the optimal parameters of similarity technique using MPA. Figure 4.7 and Figure 4.8 displays the obtained similarity between a small set of books present in Bookcrossing dataset. Similarly, we computed the similarity between items among all the items present in that particular dataset. Value '1' in the similarity index represents that the both items are same or almost same. Higher similarity index indicates high resemblance between the pair of chosen items. Low similarity index confirms that the items aren't similar at all.

For obtaining the optimal weight values of similarity index, we iterated the MPA for 100 and 200 iterations. In 100 iterations, we obtained minimal MAE value for all 6 datasets. The convergence graph for all 6 datasets used in this chapter using MPA is illustrated via Figure 4.9. In the figure, we can clearly see that the results converged towards global optima after 100 iterations in Epinions, Filmtrust and MovieLens-100K dataset. Whereas for Bookcrossing, Flixster and MovieLens-1M dataset, we iterated the optimization algorithm for 200 iterations. At this step of deciding optimal parameters, the MAE values obtained for different datasets are: Filmtrust(0.6983), Epinions (0.91624), MovieLens-100K (0.8235), Book-crossing (1.6197), MovieLens-1M (0.7924), and Flixster (0.8139).

The cost function for the optimization algorithm was the same like rating prediction equation used for predicting the rating of unknown items present in the system and then computing the error. So basically, the cost function used in the system is Mean Absolute Error (MAE) and the objective function is the proposed MMD and weight, value-based similarity technique. We tried to minimize the error (MAE) value for the ratings predicted for unknown items. The prediction of rating score for unknown items while deciding the weight value was computed using Equation 4.9.

	user_id	isbn	book_rating
0	276725	034545104X	0
1	276726	0155061224	5
2	276727	0446520802	0
3	276729	052165615X	3
4	276729	0521795028	6
...	...	...	...
2137768	276704	1563526298	9
2137769	276706	0679447156	0
2137770	276709	0515107662	10
2137771	276721	0590442449	10
2137772	276723	05162443314	8

2137773 rows × 3 columns

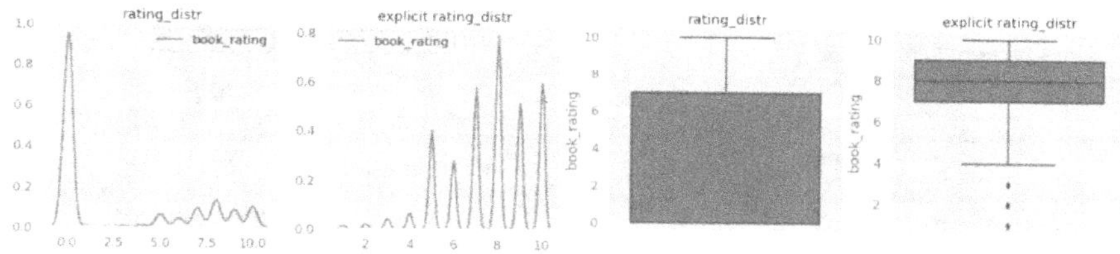

Figure 4.6 Bookcrossing - (User-Item preferences)

	034545104X	155061224	446520802	052165615X	52172137	521795028	3257224281	600570967	038550120X	342310538
034545104X	1	0.46287	0.387658	0.23741	0.789431	0.443298	0.7673	0.5683	0.51226	0.83213
155061224	0.46287	1	0.67632	0.71238	0.54972	0.32116	0.41475	0.873213	0.528118	0.71332
446520802	0.387658	0.67632	1	0.39768	0.69332	0.61787	0.47564	0.73332	0.51702	0.12885
052165615X	0.23741	0.71238	0.39768	1	0.54329	0.41871	0.61432	0.87821	0.21196	0.821332
52172137	0.789431	0.54972	0.69332	0.54329	1	0.56532	0.67291	0.11923	0.42402	0.31876
521795028	0.443298	0.32116	0.61787	0.41871	0.56532	1	0.56772	0.43102	0.65117	0.51952
3257224281	0.7673	0.41475	0.47564	0.61432	0.67291	0.56772	1	0.78645	0.56598	0.21193
600570967	0.5683	0.873213	0.73332	0.87821	0.11923	0.43102	0.78645	1	0.68712	0.78733
038550120X	0.51226	0.528118	0.51702	0.21196	0.42402	0.65117	0.56598	0.68712	1	0.11931
342310538	0.83213	0.71332	0.12885	0.821332	0.31876	0.51952	0.21193	0.78733	0.11931	1

Figure 4.7 Similarity Index Table

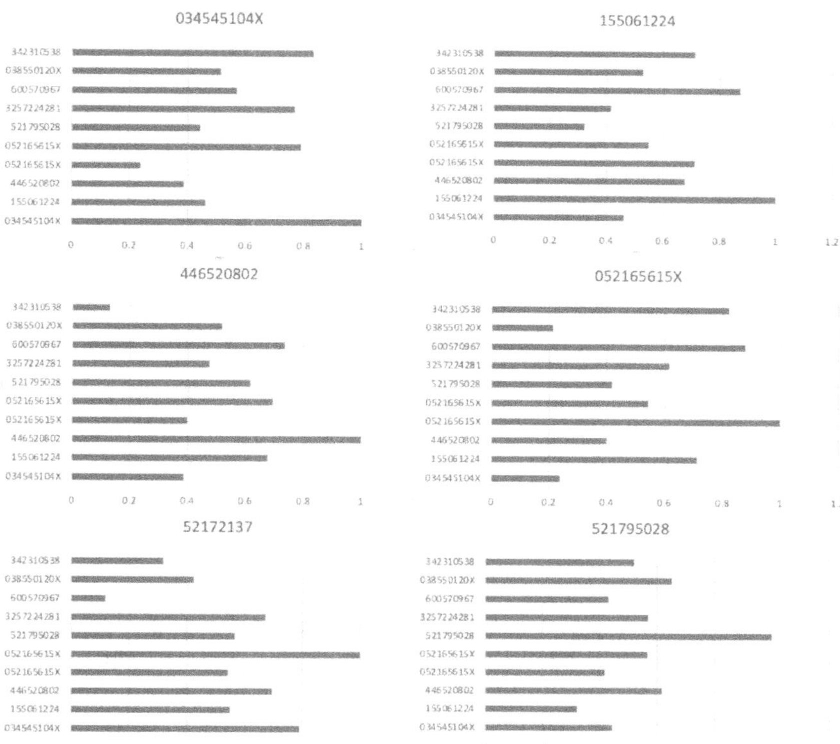

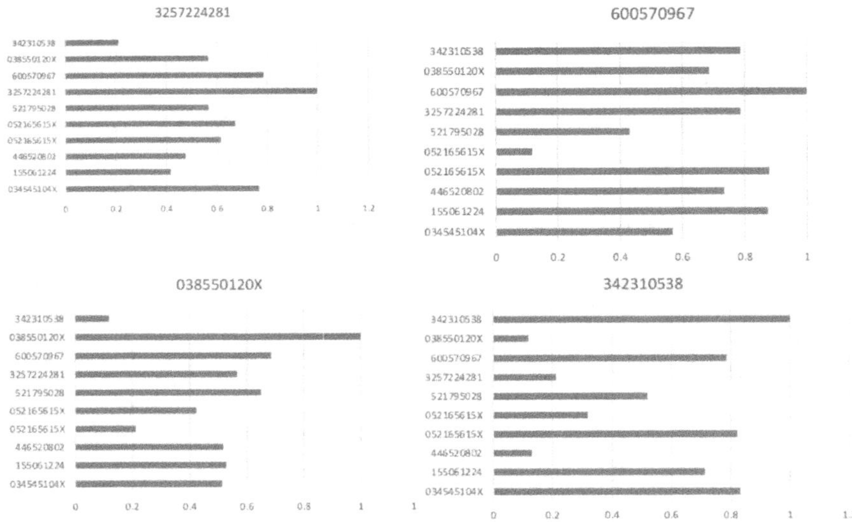

Figure 4.8 Similarity among the books (Bookcrossing dataset)

$$R'_u = \bar{R}_u + \frac{\sum_{k=1}^{K} [Sim(u, u_k) * (R_{u_k} - \bar{R}_{u_k})]}{\sum_{k=1}^{K} Sim(u, u_k)} \quad (4.9)$$

where, $R'_{u,i}$: Rating received by item 'i' from user 'u'

$\bar{R}_u$: Average rating given by user 'u'

$Sim(u, u_k)$: Similarity between two users computed using new similarity technique

R_{uk}: Rating given by user u_k to item 'i'

$\bar{R}_{u,k}$: Average rating given by user u_k

K: Total number of similar users to be considered for predicting the rating of unknown item 'i'.

We compare the obtained rating value with the actual rating and then adjust the weights of similarity technique accordingly to shift closer to the optimal weight values. After obtaining the optimal parameters of similarity technique, next step is to train the deep neural network by selecting the list of most relevant similar users with respect to any particular individual present in the system. We also explained the reasoning behind suggesting any particular movie by using an explainable component in the stacked RBMs present in the deep belief network.

We trained the deep belief network for 200 iterations. The explainable component of RBM told the target user about how many people who had similar interest like the target user gave high rating to the recommended item. Table 4.1 demonstrates a subset of explainability behind recommendation.

Table 4.1 RBMs Explainability behind recommendation

User Interest	Recommended Movie	Explanation behind recommendation
377	29	7/10 similar people rated it 3 or higher
546	1	8/10 similar people rated it 4 or higher
144	248	6/10 similar people rated it 3 or higher
88	338	5/10 similar people rated it 4
1081	154	7/10 similar people rated it 4 or higher

Table 4.1 represents the explanability for user-1 in MovieLens-100K dataset. First column (User Interest) tells us about the items in which user-1 has shown his interest by giving them high ratings. Then we generated a list of recommendation using deep belief network which had stacked RBMs in their layers. This stacked RBMs comprised of additional explainable components that helped us in informing about why any particular items is getting recommended to the user. Column-2 (Recommended Movie) enlists the item to be recommended to User-1. The third column of Table 4.1 tells us the reasoning, for e.g.: Movie-29 is recommended to User-1 because 7 out of 10 users who had similar taste like that of User-1, rated the movie-29 with 3 or higher rating score.

The MAE and RMSE value obtained by our proposed model on different datasets is discussed via Table 4.2.

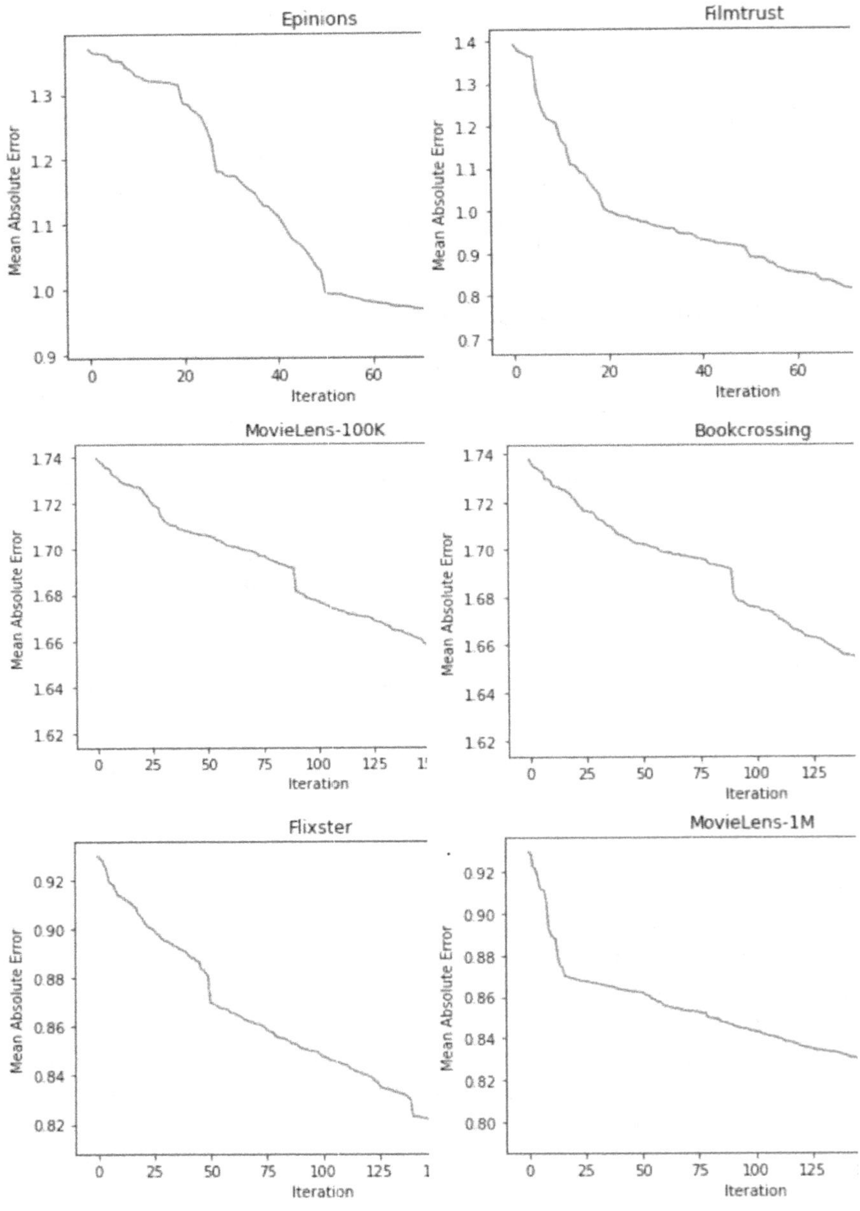

Figure 4.9 Convergence of different datasets

Table 4.2 Performance Measure of DBN model

Dataset	MAE	RMSE
Filmtrust	0.593	0.801
Epinions	0.810	0.962
ML-100K	0.71	0.743
Bookcrossing	1.47	1.96
ML-1M	0.695	0.891
Flixster	0.683	0.854

It can be clearly observed from Table 4.2 that our proposed model demonstrated good performance in most of the cases, but in few datasets like Bookcrossing, our model couldn't give low error score. But overall, the proposed model works efficiently in sparse environment without compromising the accuracy of the recommender system.

4.5 Results & Discussion

The proposed architecture based on DBN and new similarity technique yielded better results in terms of accuracy and achieved high user satisfaction as we associated explainable component along with the generated recommendation list. Although the model failed in achieving high accuracy for Bookcrossing dataset, but it achieved better recommendation accuracy for other cases. Figure 4.10 illustrates the comparative performance of our proposed model with respect to another well-known technique based on Bayesian similarity [GZY13]. It can be clearly observed from the graphical representation that the proposed model outperforms the Bayesian similarity approach. The proposed model can be further improvised by dealing with the missing values using any missing data handling approach such as matching imputation, frequent rating, or we can also use any deep learning technique such as autoencoder to fill the missing values of the dataset. This will lead to further enhancement of the accuracy of the model along with user satisfaction.

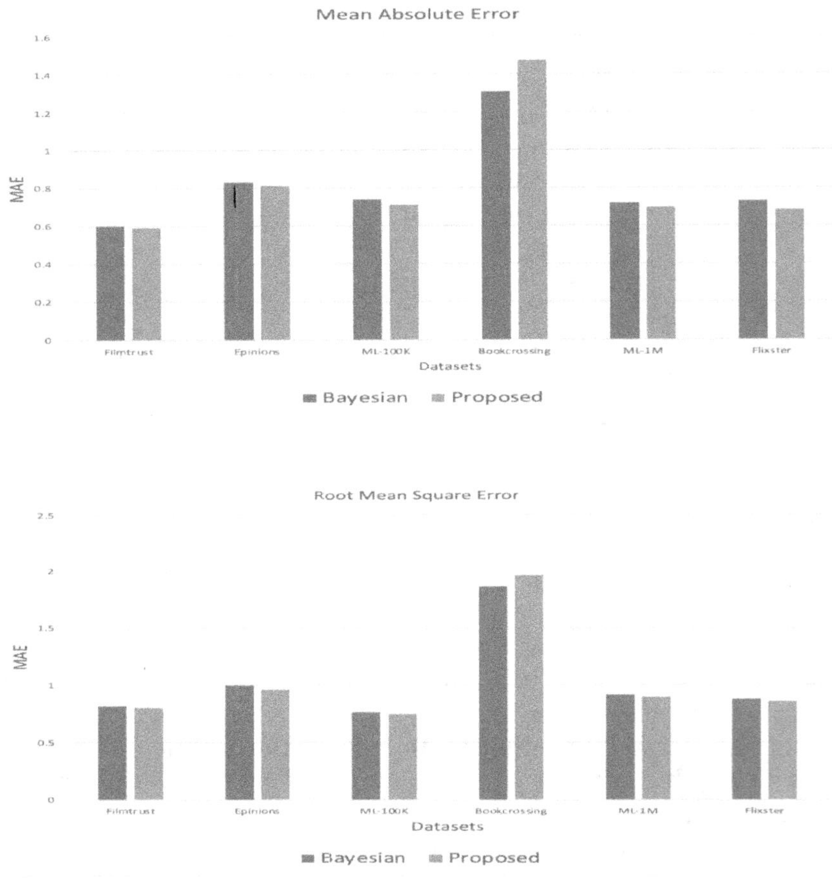

Figure 4.10 Comparative Performance of proposed DBN model

4.6 Conclusion & Future Direction

This chapter presented a collaborative filtering-based recommender system which made use of new similarity technique, an optimization algorithm, and stacked RBMs in a Deep Belief Network. The main objective of this chapter was to successfully recommend items to the users even in case of sparse environment. The chapter targeted to achieve high accuracy along with user satisfaction. This chapter presented a threefold benefit:

1. Efficiently compute similarity in case of sparse datasets.
2. Optimal parameter decision using metaheuristic approach.
3. Achieve high user satisfaction along with accuracy by providing the users with explanation behind recommending any particular item.

Although the proposed model failed badly in case of Bookcrossing dataset, in future, we can look after other approaches of deep learning to capture the rating patterns more efficiently.

CHAPTER 5

Recommendation with Deep Learning Techniques in Hybrid System

"Evidence of epistasis from hybridization studies is more scarce" – Peter R. Grant

The recommender system guides the users in a personalized manner by presenting a selective set of products from a gigantic population of credible substitutes. Over-speculation of sparse tabular data of user-item interaction draws a negative footprint on correlation modeling among the users or items present in the targeted system. In this chapter, we built an hybrid recommendation model using two layer Restricted Boltzmann Machine (RBM) and Singular Value Decomposition (SVD) technique. A metaheuristic approach namely: Spotted Hyena Optimization (SHO) is used to optimize the proposed model. We also dealt with high sparsity issue using matching imputation. Reliability of the model is tested on MovieLens (ML-100K and ML-1M) dataset and order of recommendation generated by the model reflects its supremacy over numerous state-of-art recommendation models. The predictive control structure is standardized from effectiveness on two different size datasets and can be used by the recommendation systems in case of other rating prediction datasets.

5.1 Introduction

In recent decades, the framework of recommendation has played an important role in the management of data on the dominant sectors in which consumers can determine their needs flexibly. Top-class movie advisory pages, e-commerce, online libraries, e-learning etc. makes use of several recommendation engines to examine user interests in any particular product or service. Recommender systems are the mechanisms that help individuals through the filtration of the data and by recommending products or services such as videos, news, songs, movies, etc. The key techniques used to suggest any product includes content-based filters (CBF), collaborative filters(CF) and hybrid techniques that analyzes the previous users' behavior, item characteristics, and the products or services are recommended accordingly. The key philosophy behind product suggested is to filter the products in accordance with the needs of users. The vast volume of data available on the web has made it a difficult task to locate a specific proposal in due course. The rapid growth of consumers and products on web introduced the issue of over-personalization in the recommender system. Plenitude advancements in terms of hybridized recommendation has been proposed by several researchers. However, these techniques, regardless of all progress, still need to be enhanced further to make them more relevant for real-life application implementation.

A recommending system output depends on the structured display of the user choice and item attributes. The degree of personalization is decided based on the aforementioned criteria. Construction of recommendation model is a complicated and timely task because parameters are freely adjusted while updating. Magnificent results are reflected by two efficient CF-approaches namely: Restricted Boltzmann Machine (RBM)[SMH07] and Singular Value Decomposition (SVD)[KBV09] technique. The main inclination of the chapter is towards making high accurate rating predictions as well as handling the over-speculation of sparse data in order to generate the correlation among the users and items present in the system. SVD is a matrix factorization technique that discovers the latent vectors of the users and the movies directly from the utility(user-item rating) matrix and records the user's correlation with the movie/item. The SVD provides low-rank estimations based on a significantly lower sum-squared distance. Since most of the online rating datasets are highly sparse, it produces a complicated non-convex problem, and thus a simplistic approach does not work. In this regard, a hybrid recommendation model that fuses two layer RBM model and SVD is discussed in this chapter to overcome the aforementioned challenge. RBM attains maximum likelihood by efficiently approximating the 'Contrastive Divergence' objective function. The proposed model is optimized using a metaheuristic algorithm namely: Spotted Hyena Optimization (SHO)[DK17].

The following contributions are supported by our work: *i)* Affixing stoplists to the recommendation model for avoiding offensive recommendations. *ii)* Manipulation of rating score using an equivalent matching imputation for sparsity reduction. *iii)* Use of SVD technique to provide low-rank estimations. *iv)* Implementation of a two layer RBM model with tuned hyperparameters for predicting the rating of unknown items. *v)* Administrating the model by optimizing the generation of recommendation list using SHO. *vi)* Knocking off the final order of recommendation after the categorical values for each items and release year of the items present in the recommendation list are evaluated. This chapter has the following structure: Section 2 demonstrates the techniques used by the proposed hybrid recommendation model for constructing the final set of recommendation list. Section 3 provides the main contribution and gives the final list of recommended products. In Section 4, we discussed the experimental outcomes and observations of our proposed hybrid model. We also analysed the efficiency and reliability of the proposed model using Root Mean Square Error(RMSE) and Mean Absolute Error(MAE). Final prediction order for determining the recommendation list is also explored in this section. Section 5 contrasts the outcome obtained with a variety of advanced recommendation approaches by our proposed model. Finally, on the basis of different precision measures in Section 6, we conclude our proposed model results. In this section, we also address the challenges faced by our model and the work to be done in this respect.

5.2 Experimental Methodology

Much work has been done in the area of the recommendation system with machine-learning techniques. In view of recent inquiries into recommendation strategies, the question of matrix over-speculation, data sparsity and accurate prediction is still a difficult task. In our proposed hybrid model, we used SVD technique to perform the factorization of utility matrix. Further, RBM is used to make predictions of unrated items. The parameter optimization is performed using a metaheuristic approach namely: SHO. In the closing step of the proposed model, the correct order of recommendation is decided using categorical information and release year of the items present in the generated recommendation list.

Sparsity reduction using Imputation

In our proposed model, we have used matching imputation technique to reduce the sparsity of the training utility (user-item rating) matrix. We formally minimize the sparsity of our dataset by evaluating the unrated items on basis of decided threshold value. New scores for these unrated/unspecified movies are imputed from the rating given by the highest similar user with respect to the target user for which prediction is to be made. The top similar items are decided by using cosine similarity technique. Eq. 5.1 gives the mathematical representation for matching imputation values.

$$(x_u)_r = \sum_{s=1}^{S} (x_u)_s; S \in (\text{Set of top} - \text{similar items w.r.t } 'r') \qquad (5.1)$$
$$r \& s \in T_r \text{ and } u \in U$$

where, U represents the universal set of users present in the system. $(x_u)_r$ and $(x_u)_s$ denotes the rating given to item 'r' and item 's' by selected user 'u' respectively. T_r represents the training set that comprise of item 'r' and item 's'.

Restricted Boltzmann Machine

RBM [SMH07] possess a neural network architecture having two layers namely: visible layer and hidden layer. It can be comfortably piled up to form deep net. The restriction in RBM implies that the visible and hidden layer does not include any intra-layer interactions. In RBM, one of the key drawbacks of visible layer is that it can accommodate only binary input values. Figure 5.1 illustrates the working of RBM in our proposed model. We can clearly observe that the movie ratings are represented in one-hot vector before passing it to the visible layer. For example: rating score for movie Birdman is 5 and it is represented as [0, 0, 0, 0, 1]. Similarly, the movies which are unrated are treated as blank blocks and are avoided while training the RBM. Sigmoid activation and Softmax activation function is used at visible and hidden layer respectively. Let 'H' denote the total number of hidden units such that $h_1, h_2, ...h_n \in H$. Each user 'u' present in the system has distinct RBM with distributed or shared parameters. The conditional Bernoulli distribution for modelling the hidden user features 'h' is represented by Eq. 5.2 and the modelling of binary rating matrix 'M' at visible layer is mathematically represented by Eq. 5.3.

$$P(v_i^r = \frac{1}{h}) = \frac{\exp(b_i^r + \sum_{j=1}^{H} h_j.Z_{ij}^r)}{\sum_{k=1}^{K} \exp(b_i^k + \sum_{j=1}^{H} h_j.Z_{ij}^k)} \qquad (5.2)$$

$$P(h_j = \frac{1}{M}) = \sigma(b_j + \sum_{i=1}^{m} \sum_{l=1}^{K} v_i^l . Z_{ij}^l) \quad (5.3)$$

where, 'r' denotes the rating score and v^r_i denotes the rating 'r' received by item 'i' at the visible layer 'v'. b^r_i is the bias of rating score 'r' for item 'i'. Symmetric interaction parameter between feature 'j' and rating 'r' of item 'i' is represented by Z^r_{ij}. σ is the activation function. RBM is not pliable, but Contrastive Divergence [Goo+16] helps in learning the parameters of RBM. Eq. 5.4 reflects the mathematical representation of learning RBM model.

$$\delta Z^r_{ij} = \alpha(< v^r_i h_j >_{Data} - < v^r_i h_j >_I) \quad (5.4)$$

where, α denotes the learning rate. $<v^r_i h_j>_{Data}$ determines the frequency at which item 'i' with the rating score 'r' and feature 'j' are set when the features are controlled by the observed training utility matrix effectiveness using Eq. 5.3. The distribution of the Gibbs sampling samples (Eq. 5.2, Eq. 5.3), initialized on the data, is represented by Expectation '$<v^r_i h_j>_I$' for full 'I' steps.

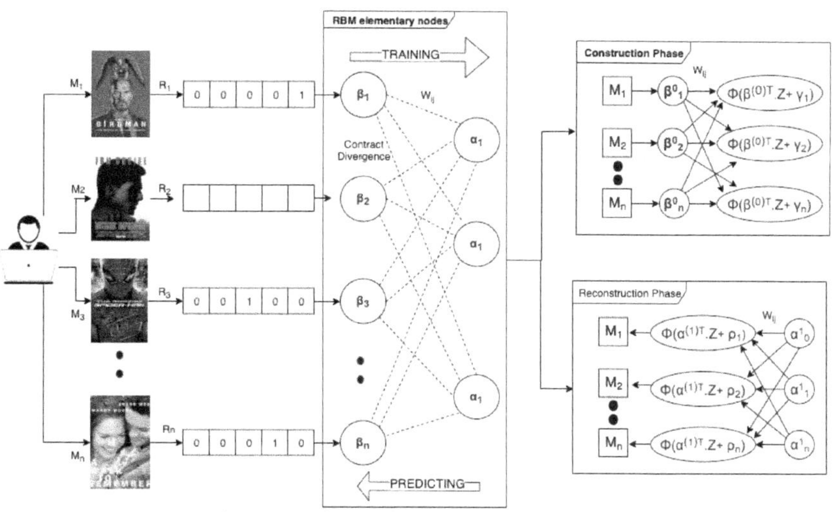

Figure 5.1 RBM based recommendation model

In the initialization step, value of I is set to '1' and is increased in the later phase as the learning converges. By incrementing the value of

'I' to a high numerical value, maximum likelihood may be randomly approximated but in reality, large 'I' values are rarely expected [CH05]. Contrastive Divergence learning has been shown to be very accurate, and the variance of learning estimates was greatly reduced [Hin02].

Singular Value Decomposition

It is a matrix factorization technique that helps in space reduction of the datasets by reducing the features from the dataset. In concern of recommender system, Singular Value Decomposition (SVD) doesn't reduce the dimensionality of rating matrix. It constructs the original rating matrix by the product of two matrices. The expected rating given to any item 'i' by user 'u' is calculated using Eq. 5.5.

$$\hat{r}_{u,i} = \mu + b_u + b_i + Q_i^T . P_u \tag{5.5}$$

where, μ denotes the average rating received by all the items present in the system. b_u represents the average rating given by user 'u' - μ and b_i is the average rating received by item 'i' - μ. P_u and Q_i denotes the user and item vector respectively. The value of b_u and P_u is kept as '0' for unknown users whereas the value of b_i and Q_i is kept as '0' for unknown items. Figure 5.2 illustrates the working principle of SVD in recommender systems. In order to estimate all the unknowns, present in the system, regularized squared error represented by Eq. 5.6 is minimized.

$$\text{minimum}(P, Q, b_u, b_i) \sum_{r_{u,i} \in R_{train}} (r_{u,i} - \hat{r}_{u,i})^2 + \alpha(b_u^2 + b_i^2 + ||P_u||^2 + ||Q_i||^2) \tag{5.6}$$

where, α denotes the regularization term. Stochastic gradient descent is used to perform the minimization of Eq. 5.6 and the mathematical representation of minimization steps are given by Eq. 5.7.

$$\begin{aligned} b_u &\leftarrow b_u + \beta(E_{u,i} - \alpha.b_u) \\ b_i &\leftarrow b_i + \beta(E_{u,i} - \alpha.b_i) \\ P_u &\leftarrow P_u + \beta(E_{u,i}.Q_i - \alpha.P_u) \\ Q_i &\leftarrow Q_i + \beta(E_{u,i}.P_u - \alpha.Q_i) \end{aligned} \tag{5.7}$$

where, β represents the learning rate and $E_{u,i}$ is equivalent to the difference between actual and predicted rating. The above steps are repeated for 'n' epochs on all the ratings data available on training set.

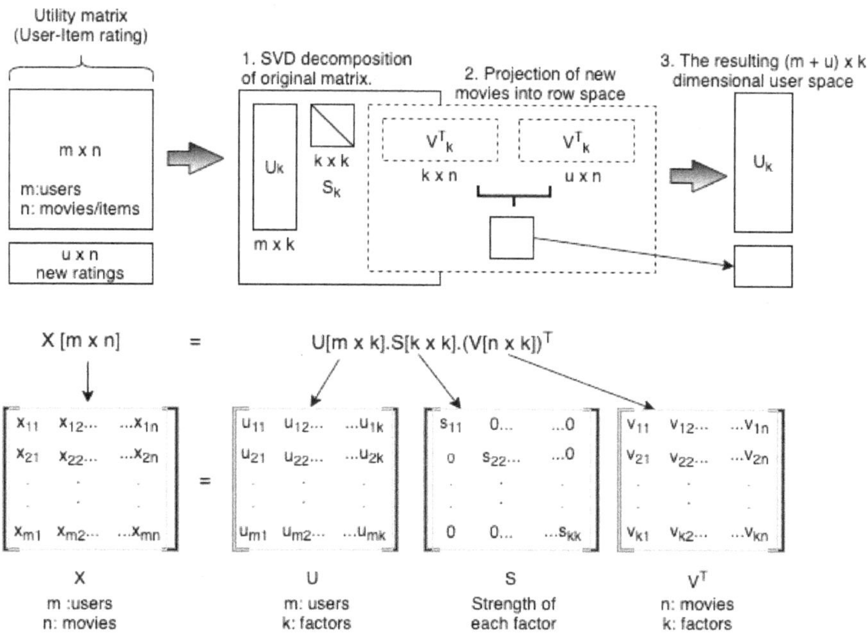

Figure 5.2 SVD based recommendation model

Spotted Hyena Optimization

The main inspiration behind using Spotted Hyena Optimization (SHO) is the social relation among the hyenas. Spotted hyenas are expert hunters of three other species (i.e., aardwolf, brown and striped). The most popular ones are spotted hyenas. The hunting behaviour of spotted hyenas is illustrated using Figure 5.3. Spotted hyenas typically reside and hunt in groups. They have more than 100 participants forming a network of trustworthy mates [IBH15]. In order to increase their network, they bind themselves to another spotted hyena who is either a mutual friend or have some kind of parentship relation. They hardly link to any anonymous hyena to form groups. Spotted hyena are social species that can interact with each other through specific calls like gestures and signals. Figure 5.3 demonstrates the searching, pursuing, encircling and ambushing the prey in a grouped manner. The techniques of hunting and the social connection between spotted hyenas are modelled mathematically to establish and optimize SHOs.

Figure 5.3 Spotted hyenas attacking its prey

Encircling the target prey

Just because the search space is not known to us in prior step, we consider the best candidate solution available to us as the target prey. The mathematical model addressing the behaviour of other search agents is explained by Eq. 5.8.

$$
\begin{aligned}
Dist_h &= |\vec{C_1}.\vec{P_r}(i) - \vec{H}(i)| \\
\vec{H}(i+1) &= \vec{P_r}(i) - \vec{C_2}.Dist_h \\
\vec{C_1} &= 2.\vec{r_1} \\
\vec{C_2} &= 2.\vec{h}.\vec{r_2} - \vec{h} \\
\vec{h} = 5 - (iter * (5/M)); iter &= 1, 2, 3..M
\end{aligned}
\tag{5.8}
$$

where, $Dist_h$ represents the distance between the spotted hyena and target prey.

$\vec{C}_1$ and $\vec{C}_2$ are the co-efficient vectors. 'i' denotes the current iteration. $\vec{H}_1$ and $\vec{P}_r$ tells about the current position of the hyena and current position of target prey respectively. $\vec{r}_1$ and $\vec{r}_2$ are two random vectors that lie in range of [0,1]. Figure 5.4 illustrates the effect of $Dist_h$ and $\vec{H}$ (i + 1) in a 2-D surrounding. In Figure 5.4, (X, Y) denotes the position of spotted hyena and the position of target prey is denoted by (X', Y'). A certain number of positions are possible to be reached from the current location after adjusting the value of $\vec{C}_1$ and $\vec{C}_2$.

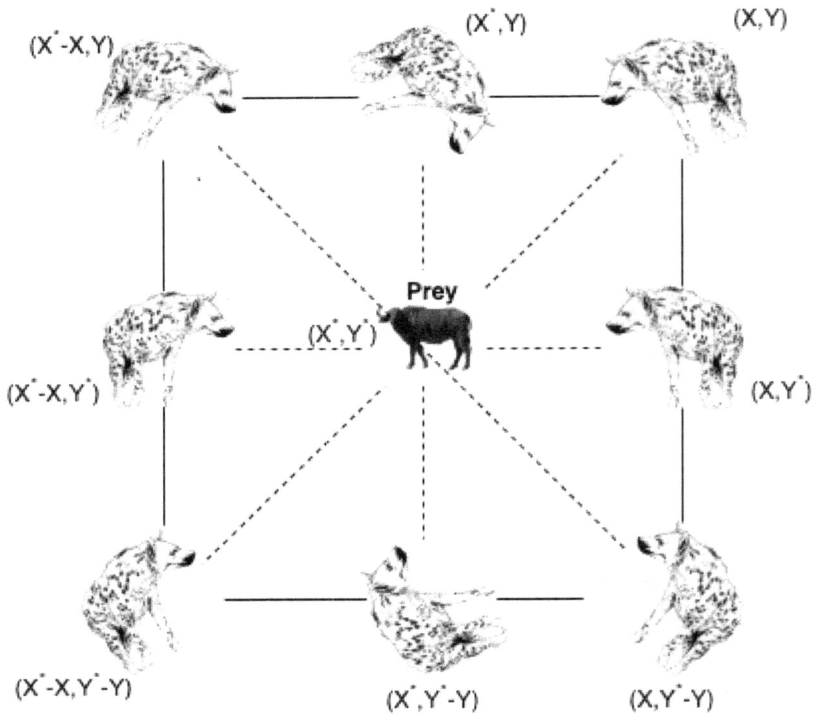

Figure 5.4 Positions of hyenas in 2-D space

Chasing the target prey

As discussed in earlier section about the hunting behaviour of spotted hyenas, they usually attack in groups/clusters. The formed cluster comprises of trusted network. The location of the target prey is known by the best search agent. The other agents present in the cluster update their positions based on this best solution. Mathematically it is represented by Eq. 5.9.

$$\begin{aligned} Dist_h &= |\vec{C_1}.\vec{B_h} - \vec{O_k}| \\ \vec{O_k} &= \vec{B_h} - \vec{C_2}.Dist_h \\ \vec{C_h} &= \vec{O_k} + \vec{O}_{k+1} + \vec{O}_{k+2} + \ldots + (\vec{B}_{k+N}) \end{aligned} \tag{5.9}$$

where, $\vec{B}_h$ represents the best spotted hyena in the initial step. ϑ_k denotes the vector position of other spotted hyenas present in the network. $\vec{C}_h$ forms a cluster of 'N' number of optimal solution present in the population.

Exploitation of search space (Attacking the target prey)

The mathematical model that discusses the attacking behaviour of spotted hyenas is given by Eq. 5.10. With the help of this equation, the spotted hyenas update their position while keeping the best obtained solution as the base point of reference. The search agents update their position in such way that they move towards the target prey.

$$\vec{H}(i+1) = \frac{\vec{C_h}}{N} \tag{5.10}$$

Exploration of search space (Searching for the target prey)

In this step, group of hyenas look for available prey in the nearby surrounding. In terms of optimization, we consider it as looking for possible solution in the population. The process of searching for prey is termed as "Exploration". If |C2| > 1, the system acts in the direction of Exploration whereas, if |C2| < 1, the system acts in the direction of Exploitation. Figure 5.5 illustrates the behaviour of spotted hyenas in terms of Exploitation and Exploration.

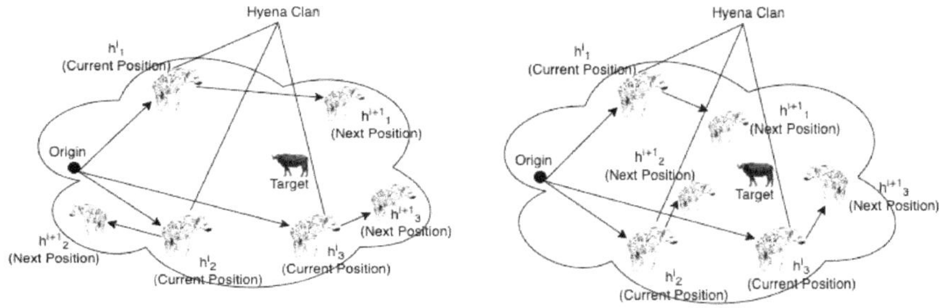

Figure 5.5 Searching and attacking prey

Performance measures

Various reliable tests have been used to verify that the proposed hybrid model and the fused optimization technique work better than a variety of cutting-edge customization methods. Root Mean Square Error (RMSE) and Mean Absolute Error (MAE) are the main test metrics for measuring our model results. Less RMSE and MAE reflect high accuracy efficiencies. In a series of forecasts, the MAE and the RMSE can be used to determine the variability for errors. The RMSE MAE value; the bigger the difference among the different sample errors, the greater the variance. All residuals are of the same magnitude if the value of RMSE = MAE.

Mean Absolute Error

The MAE calculates, without taking into account their direction, the average size of errors in a set of predictions. It calculates the accuracy of the model that yields continuous variables. The absolute difference between the forecast value and the corresponding observation is averaged over the test samples. The MAE is a linear ranking, meaning that deviations are weighted on average equally. MAE is represented by Eq. 5.11.

$$MAE = \frac{1}{N}\sum_{i=1}^{N}|Y_i - A_i| \quad (5.11)$$

where, Y_i and A_i represent the predicted and actual value respectively. 'N' denotes the total number of observations.

Root Mean Square Error

In RMSE, the difference between estimates and the corresponding observed values is squared each time and the sample is then summed. In the final step, average is square-rooted. The RMSE gives large failures a relatively high weight as the errors is squared before they are averaged. The RMSE is therefore most beneficial when significantly large errors are particularly expected in the prediction model. Eq. 5.12 depicts the mathematical form for RMSE calculation.

$$RMSE = \sqrt{\frac{\sum_{i=1}^{N}(Y_i - A_i)^2}{N}} \quad (5.12)$$

where, Y_i and A_i represent the predicted and actual value respectively. 'N' denotes the total number of observations.

5.3 The Proposed Approach

This section demonstrates several stages involved in the construction of our proposed model as shown in Figure 5.6. The model starts with construction of efficient utility matrix, followed by splitting of formed utility matrix into 75-25 train-test ratio. Before training the proposed model using the training set, stoplist is applied to trim out the items which might lead to give offensive recommendations. The main motive behind blending of stoplist in our proposed model is to avoid unwanted controversy in real world scenario. Imputation technique is used to deal with sparsity of the dataset. In further step, we have used RBM and SVD technique to make collaborative decision on rating prediction for any set of unrated items present in the system. The optimal weightage given to both the techniques play a crucial role in increasing the accuracy of the hybrid model. The optimal weight values are calculated using SHO. Accuracy of the model has been tested using RMSE and MAE performance metric. The final order of recommendation is decided on basis of the item release year and the predicted rating for that item. Detailed description of each stage is given in the forthcoming subsection.

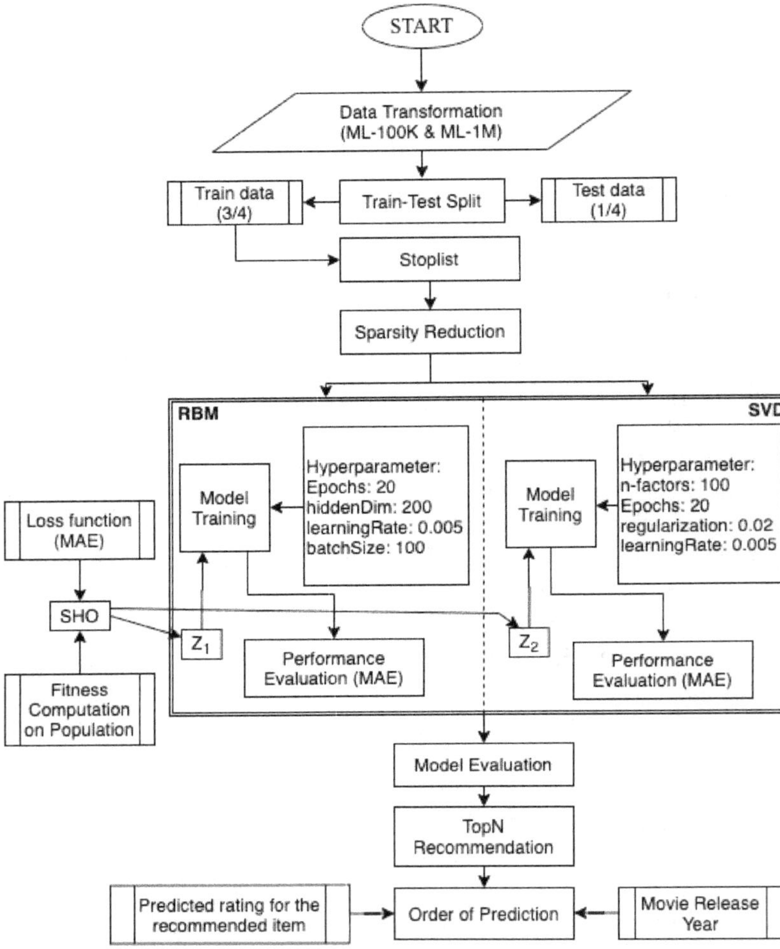

Figure 5.6 Proposed hybrid recommendation model

Data Transformation

The MovieLens dataset has been used, which contains movie ratings in scale of 1 to 5, where 1 represents low rating and 5 is high rating. Minimum of 20 movies are rated by each user present in the system. In the first stage of proposed model, we performed transformation of dataset in order to form a utility (user-item rating) matrix such that each movie are placed in separate rows and each users are placed in separate columns of the matrix. The unrated movies are represented by NA (Not Available) in the utility matrix. Table 5.1 represents the description of the dataset being used in the chapter.

Table 5.1 Dataset description

Dataset	#Users	#Items	#Ratings	Sparsity (%)
ML-100K	943	1682	1,00,000	93.7%
ML-1M	6040	3900	10,00,209	95.8%

Stoplist

It checks whether the recommendation might cause unwanted controversy in a real-world environment. It helps in preventing the system from giving offensive recommendations. Following is the set of item attributes that can be added in stoplist: Adult-oriented content, vulgarity, legally prohibited topics, terrorism/political extremism, bereavement/medical, competing products, drug use, religion, etc. In our proposed model, we have tried to avoid the recommendation of adult-oriented content, drug use and religion-oriented items to any user present in the system.

Sparsity reduction using Matching Imputation

In this section, we performed imputation of missing values in order to cutback sparsity of the training utility matrix upto some extent. Firstly, we search for the top similar movies with respect to movie for which we want to predict the rating. Further down the line, we impute the rating of the highest similar movie to the unrated movie for selected specific similar users. In such way, we reduce the sparsity without compromising the performance of model. If the topmost similar items are also unrated, we leave the unrated item rating as NA (Not Available).

Model training and rating prediction using RBM and SVD

We make a collaborative decision on the rating value of unrated items by using RBM and SVD technique. The weightage given to both the techniques in taking the final decision on rating value is decided by SHO. RBM tries to reconstruct the rating of items (user-item) pair for which we don't know the rating yet. At the visible layer, visible nodes intakes rating given by specific user to all the items. In order to make the RBM generalized for any rating scale, the possible set of ratings are converted into one hot encoded format. This one hot encoding ensures that the system doesn't assumes high rating as more important as compared to low rating values. The weight and bias plays key role in predicting the missing ratings. Since most of the ratings in the movie dataset is unknown, we exclude the unrated items while training the RBM. Softmax is used to regenerate the final rating. In the movie-recommendation system, we pass the 'k' users with their rating scores into the RBM neural network and perform training for certain number of epochs. One epoch denotes one complete iteration of training the model. Each input 'i' that is transferred through the neural network reflects a single consumer ranking for all 'n' items present in the system. The links between the visible layer and the hidden layer are used for learning the weights. Addition of bias values at the hidden layer ensures that some of neuron's fire. The formed equation (W I + B), where 'W' denotes weight and 'B' denotes Bias is passed through the activation function. The hidden layer activation results in stochastically generated final outputs. This randomness in the system helps to establish a more efficient and stable generative model. The RBM iterates forward and backward to learn how to best weigh when trying to build a robust generative model using Gibbs Sampling [Wig+19] approach. By iteratively adjusting the weights of the neural net, the RBM will find the connections between the input functions and decide the factors that are most important. RBM learns to maximize the approximation of the original data. This learned distribution of probabilities helps RBMs to predict unprecedented results. The RBM attempts to forecast item ratings that the user has never seen based on correlations between the targeted user and other users and on the values that the other items received from other correlated users. SVD constructs the original rating matrix by the product of two matrices. The ratings for unknown items are predicted using Eq. 5.5.

Final list of recommendation

After deciding the Top-N items for performing recommendation, the final list of recommendation is generated on basis of the item release year and the rating predicted by the system for that item. If we recommend part-2 of any movie, followed by part-4 and part-3, the ordering will not be appreciated by the end user and thus proper ordering of movies comes into picture. In our proposed model, we rearranged the recommendation list based on item categorical information and item release year.

5.4 Experimental Outcomes and Observation

This section of the chapter discusses the experimental outcomes of each stage of the proposed model and the observation drawn from each stage of model is also analysed. After the dataset is pre-processed, we blend stoplist on the training utility matrix. Before starting the training of proposed model, the items which are related to adult-content, drug use or religion oriented are blocked by the system. Figure 5.7 illustrates the blocked items before training of model actually begins. The matching imputation helps in reducing the sparsity of ML-100K and ML-1M dataset from 93.7% and 95.8% to 62.5% and 67.2% respectively.

The experimental outcome obtained after implementing SVD technique on ML-100K dataset is reflected by Figure 5.8. The recommendation result shown in Figure 5.8 is for User ID - 85. Likewise, recommendation list for other users can be generated by SVD technique. We have made recommendation of Top-10 relevant items to any targeted user. Similarly, the recommendation list and accuracy obtained using RBM technique is illustrated by Figure 5.9. In order to avoid the negative speculation of sparse dataset, we combine the aforementioned techniques. The weightage given to both the techniques while taking a collaborative decision on rating prediction is taken by SHO. The result obtained using hybrid approach outperforms the standalone RBM and SVD technique. Figure 5.10 demonstrates the results obtained using proposed hybrid model. The average RMSE and MAE value obtained using proposed model is 0.8239 and 0.6281 respectively. The list shown at the bottom of Figure 5.8, Figure 5.9 and Figure 5.10 reflects the items recommendation for targeted USER ID '85'. The optimal weight value is decided by SHO and it lies in the range of [0,1], where '1' denotes high weightage and '0' represents negligible contribution in taking decision.

```
Using recommender  Hybrid

Building recommendation model...
Blocked  drugstore cowboy (1989)
Blocked  love & sex (2000)
Blocked  sex, lies, and videotape (1989)
Blocked  sexy beast (2000)
Blocked  sex and lucia (lucãa y el sexo) (2001)
Blocked  sex tape (2014)
Blocked  everything you always wanted to know about sex * but were afraid to ask (1972)
Blocked  opposite of sex, the (1998)
Blocked  love and other drugs (2010)
Blocked  sex: the annabel chong story (1999)
Blocked  sex drive (2008)
Blocked  sexmission (seksmisja) (1984)
Blocked  sex and the city (2008)
Blocked  talking about sex (1994)
Blocked  relax... it's just sex (1998)
Blocked  loss of sexual innocence, the (1999)
Blocked  midsummer night's sex comedy, a (1982)
Blocked  sex and the city 2 (2010)
Blocked  sex ed (2014)
Blocked  the opposite sex (2014)
```

Figure 5.7 Blocked Items – Stoplist

```
Loading movie ratings...

Computing movie popularity ranks so we can measure novelty later...
Estimating biases using als...
Computing the cosine similarity matrix...
Done computing similarity matrix.
Evaluating  SVD - Untuned ...
Evaluating accuracy...
Analysis complete.

Algorithm  RMSE      MAE
SVD - Tuned 0.9039    0.6984

Legend:
RMSE:     Root Mean Squared Error. Lower values mean better accuracy.
MAE:      Mean Absolute Error. Lower values mean better accuracy.

Using recommender  SVD - Tuned

Building recommendation model...
Computing recommendations...

We recommend:
Lord of the Rings: The Fellowship of the Ring, The (2001) 4.575751068047649
Howl's Moving Castle (Hauru no ugoku shiro) (2004) 4.486299465683202
Ideal Husband, An (1999) 4.47074877424001
African Queen, The (1951) 4.411259849278635
Raiders of the Lost Ark (Indiana Jones and the Raiders of the Lost Ark) (1981)
4.399824892559434
Usual Suspects, The (1995) 4.399449909291453
Eternal Sunshine of the Spotless Mind (2004) 4.395973587119041
Spirited Away (Sen to Chihiro no kamikakushi) (2001) 4.394183044899286
Fargo (1996) 4.389889084419128
Searching for Bobby Fischer (1993) 4.351505720257753
```

Figure 5.8 Recommendation list generated using SVD

```
Evaluating RBM - Tuned ...
Evaluating accuracy...
Analysis complete.

Algorithm  RMSE     MAE
RBM -Tuned 1.1896   0.9934

Legend:
RMSE:    Root Mean Squared Error. Lower values mean better accuracy.
MAE:     Mean Absolute Error. Lower values mean better accuracy.

Using recommender RBM
Building recommendation model...
Computing recommendations...

We recommend:
Howl's Moving Castle (Hauru no ugoku shiro) (2004) 3.3278856
Harry Potter and the Deathly Hallows: Part 2 (2011) 3.3252628
Hurt Locker, The (2008) 3.3100324
Downfall (Untergang, Der) (2004) 3.3088908
Harry Potter and the Half-Blood Prince (2009) 3.3071156
Amores Perros (Love's a Bitch) (2000) 3.3010027
My Neighbor Totoro (Tonari no Totoro) (1988) 3.2987766
Moon (2009) 3.2978768
There Will Be Blood (2007) 3.294481
Blood Diamond (2006) 3.2944016
```

Figure 5.9 Recommendation list generated using RBM

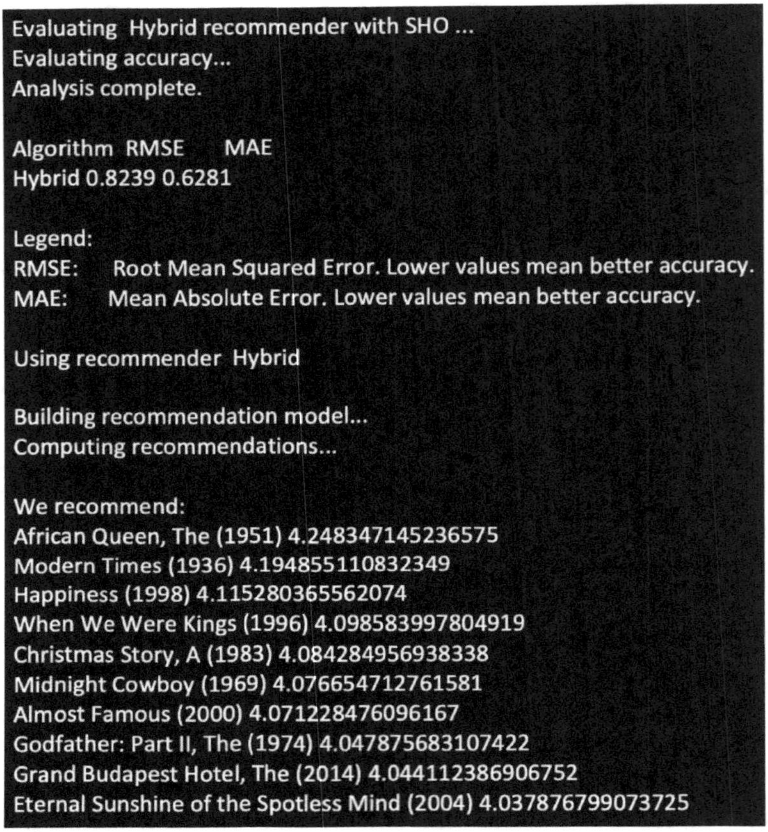

Figure 5.10 Recommendation list generated using proposed hybrid model

The prediction order in the final phase of the proposed model is determined by the predicted rating and item categorical information (Genres), as shown in Table 5.2. We group the items based on their categorical information. The MovieLens dataset comprise of 19 categories in total. Firstly, we arrange the recommended list of items in decreasing order of predicted rating, then we group the same category item together and rearrange them on basis of release year. If two items belong to same category, arrangement in the decreasing order of release year will ensure the model from recommending the correct sequence of items. E.g.: Harry potter - III will be recommended before recommending Harry Potter -II and Harry Potter -I. Our model gave promising results as compared to numerous comparative recommendation models as discussed in the upcoming section.

Table 5.2 Order of recommendation

Movie Name	Release Year	Category	Predicted Rating
Africa Queen, The	1951	Action, Adv., Rom., War	4.24
Happiness	1998	Comedy	4.11
Modern Times	1936	Comedy	4.19
When We Were Kings	1996	Documentary	4.09
Grand Budapest Hotel, The	2014	Comedy, Drama	4.04
Almost Famous	2000	Comedy, Drama	4.07
Christmas Story, A	1983	Comedy, Drama	4.08
Midnight Cowboy	1969	Drama	4.07
Godfather: Part II, The	1974	Action, Crime, Drama	4.04
Eternal Sunshine of the	2004	Drama, Romance, Sci-Fi	4.03

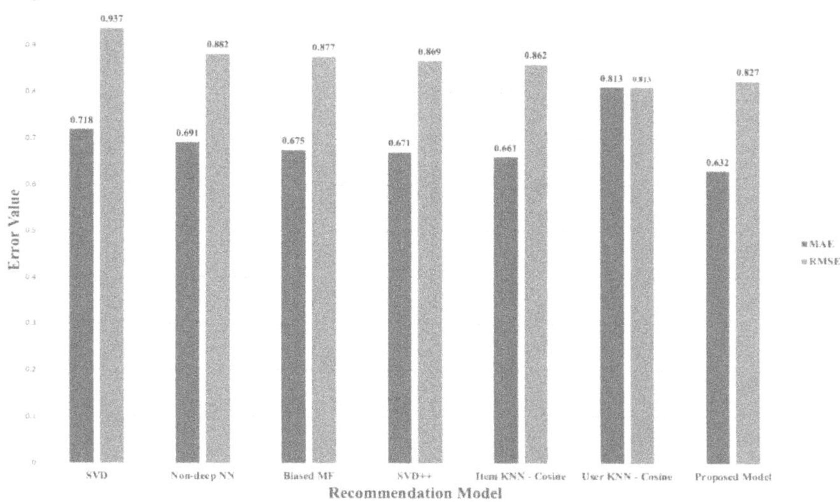

Figure 5.11 Model Comparison - MovieLens-100K

5.5 Results and Discussion

By comparing RMSE and MAE with a number of state-of-the-art comparative recommendation models, we display our findings through Figure 5.11 and Figure 5.12. Compared to existing methods namely: SVD [KKB20], Non-deep NN[KKB20], Biased MF[KKB20], SVD++[KKB20], Item KNN - Cosine[KKB20], User KNN - Cosine[KKB20], DNNRec [KKB20], GA & Gravitational Emulation[Moh+19], Collaborative Deep Learning [WWY15], RippleNet [Wan+18a], U-AutoRec [Sed+15], I-AutoRec [Sed+15], NNMF [DR15], LLORMA-Local [Lee+13], LLORMA-Global [Lee+13], U-CF-NADE-S(2 layers) [Zhe+16] and I-CF-NADE-S(2 layers) [Zhe+16], we notice that the accuracy of our proposed model is far better. The obtained RMSE and MAE value on MovieLens 100K dataset and MovieLens 1M dataset shows that the proposed hybrid model along with Spotted Hyena Optimization yields high accuracy as compared to several other recommendation techniques. In addition to accuracy of the recommendation model, we found that the item attributes and predicted rating proved to be few important parameters for deciding the final list of recommendation.

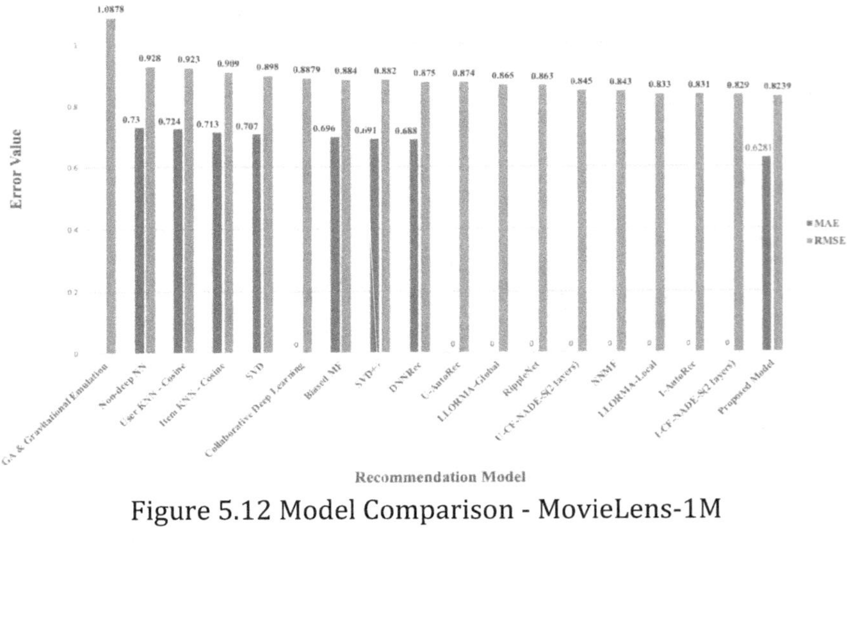

Figure 5.12 Model Comparison - MovieLens-1M

5.6 Conclusion and Future Work

This chapter presented a hybrid recommendation framework based on SVD, RBM and SHO technique. The suggested approach leverages a dynamic model with optimization technique, to learn non-linear latent factors for users and items, blends learning capabilities with side knowledge accessible on users and items to build a hybrid framework. The effects of proposed model is contrasted with a number of heuristic, next-generation, machine-based and matrix factorizing techniques. The efficacy of proposed model is demonstrated using MovieLens benchmark dataset. The chapter has five-fold contributions: It addresses the negative effect of sparse tabular data on correlation modeling using matching imputation, affixing stoplists to the recommendation model for avoiding offensive recommendations, providing low rank estimations using SVD and RBM technique, administration of model by optimizing the generation of recommendation list using SHO and knocking of correct order of prediction using item categorical information and item release year. One of the drawbacks of the proposed work is that it deals with the rating prediction issue of the recommender systems rather than the ranking case or streaming scenario. In future, we aspire to focus in this direction. We also intend to improve the proposed model using following facets: a) Temporal relationship between user and item based on history. b Usage of recurring neural networks (RNN) to leverage the sequential nature of the predictions.

CHAPTER 6

Recommendation in Context-Aware System with Similarity Technique Based on Feature Weight Learning

"Without context, a piece of information is just a dot. It floats in your brain with a lot of other dots and doesn't mean a damn thing. Knowledge is information-in-context connecting the dots" — Michael Ventura

The recommender framework personalizes consumers by offering a targeted variety of items from a vast community of believable alternatives. This knowledge filtering is done through evaluating user or item similarities by the most common collaborative filtering process. In this chapter, we presented a similarity metric that includes weights and values. Values are determined by taking into consideration the matching consumer set on which similarity needs to be calculated. The optimal weight values are determined by an improved Crow Search Algorithm (iCSA). The stability of the exploitation and exploration of CSA is enhanced by the use of three factors: Levy flight, accustom of commutable operator and proceedings component. On Jester, MovieLens 100K, MovieLens 1M, MovieLens 10M and MovieLens 20M the efficiency of the proposed model is validated. The comparative observation indicates that the augmented model performance is better than several state-of- art personalized recommender systems. The proposed model is less sensitive to dataset dimensions and shows outstanding progression in terms of accuracy. The generated recommendation list is made context-aware by monitoring the items in which the user has shown interest recently.

6.1 Introduction

Over the last decade or two, the recommender framework has played a major role in controlling the proliferation of information on domains where consumers have the flexibility to rank their preference [MTF20]. Consumers frequently interact with Recommendation Systems (RS) by using digital platforms and software, from Walmart.com to Netflix and media sites. RS makes use of algorithms that recommend what a user wants, such as a particular video. More technically, there are functions which take processed data about a users interests (e.g. movies) and forecast how a user rate the items under consideration (e.g. new movies) and how they rate a group of items individually or as a collection. In this paper, we will discuss more about the essence of recommendations and role of similarity technique in performing filtration of information, but even this basic overview is enough to make it clear that recommendation systems capture, curates, and uses a large number of personal data to run effectively and efficiently. They ultimately mould human understanding with interactive worlds and social experiences.

The key philosophy behind product suggested is to filter the products in accordance with the needs of users. The vast volume of data available on the web has made it a difficult task to locate a specific proposal in due course. The rapid growth of consumers and products on web introduced the issue of over-personalization in the recommender system. Plenitude advancements in terms of hybridized recommendation has been proposed by several researchers. However, these techniques, regardless of all progress, still need to be enhanced further to make them more relevant for real-life application implementation. The foundation of any RS is its algorithm for filtering information [SD19b]. The most widely used collaborative filtering system computes the similarity among the users using either distance-based or correlation-based techniques. The key motivation behind designing a new similarity technique based on weight and values are as follows:

1. Low correlation because of the identical scores offered to common products by the two consumers. This problem occurred primarily in the best-known Pearson correlation coefficient (PCC) and also in the Squared PCC.
2. High correlation regardless of the disparity in user rating behaviours against popular products. This issue persisted primarily with Cosine, PCC and SPCC. Adjusted Cosine has demonstrated efficacy for this case.
3. If the proportion of common scores is not recognised, the outcomes are unreliable. Mean Squared Deviation (MSD) measures only the average difference between users, but it does not take into consideration the typical ratings ratio. This leads to low precision.

4. The elimination of the absolute values of commonly rated products would make it impossible to discriminate between different consumers. The Jaccard similarity is based purely on the ratio of the common score and does not take into consideration the absolute measure of the score. This makes the distinction between users difficult to determine.

Collaborative filtering, based on user interests, informs consumers of products. Therefore, a record of user preference history must be available. However, the database is often very limited, i.e., the customer only scores a very small number of items. So far, several researchers have focused on the precision of forecasts and proposed solutions to this-mentioned challenge. Table 6.1 demonstrates the work done by several researchers in this direction.

Table 6.1 Traditional similarity technique performance

Ref.	**Similarity Technique**	**Description & Limitation**
[JE09]	Trust Based Similarity Network	They suggested a metric of similarity based on the sigmoid function. One of the key demerits of the metric is its poor behaviour in the measurement of similarities with a small group of frequently valued products.
[Ahn08]	PIP (Proximity - Impact - Popularity)	They introduced a new heuristic measure of similarity which focuses on improving the recommendation efficiency in cold-start conditions, in which only a tiny proportion of rating scores are available for each consumer. However, this measure takes local details only into consideration and does not take account of the global choice of customer reviews.
[BOH12]	Singularity based similarity	They took the assumption that the findings obtained by standard similarity tests could be improved by taking qualitative data into account.

[Bob+12]	Significance based similarity	The methodology was used to measure the importance of a product, the significance of a user and the significance of a product for a user to recommend for other users.

Apart from the similarity techniques mentioned in Table 6.1, few other traditional similarity techniques being widely used by recommender system includes: Cosine Similarity, Pearson Correlation Coefficient (PCC), Jaccard Similarity, Mean Square Difference (MSD) and Mean Jaccard Difference (MJD). PCC does not consider the size of the consumer set. To address this problem, a weighted PCC [Her+17] was proposed taking the assumption of the trust of the neighbour into consideration. MJD offered way to solve the problem of cold user problem. This approach takes place in three stages: first, to pick measures of similarity; following this step, this metric includes six measures for similarity. And neural network learning assesses the weight of each measure of similarity. In end, the projection can be done as per the new metric. In addition to the use of numerical ratings information (via Jaccard), JMSD [BSB10] improvises recommendation accuracy to some degree, and also improvises the consistency of ratings with numerical information (via MSD). In addition to the traditional similarity measures, some researchers have proposed fuzzy recommendation frameworks [Hu+15] [MJ16] to approximate the customer-to- product relationship. The focus is on consumers, products and reviews for fuzzy accounts in particular. It contrasts those profiles to assess noise preference by measuring noise levels using fuzzy profiles [YM17].

After examining the conventional similarity measures and their variations, we find that most customers rate very limited items in the framework. In the end this tends to less precise values of resemblance. This motivated us to create a new similarity metric with a substantial functionality benefit and a wide variety of applications. It can be introduced without any specific details in virtually all current recommender schemes. Few most widely used optimization techniques in the domain of recommender frameworks includes[MKI19]: Artificial Bee Colony (ABC), Bat Algorithm (BA), Cuckoo Algorithm (CA), Genetic Algorithm, Gravitational Search Algorithm (GSA), Grey Wolf Optimizer (GWO), Invasive Weed Optimization (IWO), and Particle Swarm Optimization. We have used different nature inspired meta-heuristic approaches and gradient based approach such as Gradient Based Optimizer (GBO) to determine the optimal weight values of the proposed similarity metric. The blueprint of proposed similarity metric is discussed in Section II of the paper. A user-friendly similarity measure, assisted by an effective meta-heuristic approach, is the main guiding force behind the paper to achieve high precision and less complicated estimation. This paper work includes the following major contributions:

1. Estimation of user similarity value using new similarity metric.
2. Deciding the top relevant neighbours using relative similarity index (RSI).
3. Administrating the optimal weights obtained using different nature inspired meta-heuristic approaches such as GA, PSO, CSA and gradient based approach such as GBO.
4. Implementation of an upgraded CSA that incorporates three additional factors in CSA namely: Levy flight, accustom of commutable operator and proceedings component. The improved CSA (iCSA) achieved better performance in terms of accuracy and convergence.
5. Analyse the current prediction context and generate the list of items to be recommended accordingly.

In Section II, we present the similarity metric based on feature weight learning along with relative similarity index calculation. Section III discusses the nature inspired meta-heuristic approaches and gradient based approach used in our work for optimizing the weight values of the proposed similarity metric. The flow of proposed model is discussed in Section IV and the experimental setup and outcomes are illustrated in section V of the paper. Section VI and Section VII highlights the obtained results and conclusion drawn from the research work.

6.2 Similarity Metric based on Feature Weight Learning

The Vector Space Model (VSM) is the foundation for most of the classification approaches. In order to compute the similarity between two arbitrary vector 'X' and 'Y', several similarity metrics: Cosine, PCC, Euclidean, etc. can be used. For e.g.: Euclidean distance can be computed by using Equation 6.1.

$$\text{dist}(X_a, Y_b) = \sum_{i=1}^{n} (X_{ai} - Y_{bi})^2 \tag{6.1}$$

In Equation 6.1, distance usually refers to all characteristics and assumes that they all have the same distance effects. These similarity measures do not take into account the relation of characteristics, which leads to inaccurate distances and then on to projections. The similarity metric would be deceptive relative to all the attributes of equivalent weight values. A large number of irrelevant attributes will dominate the distance. This difficulty that arises when there are many irrelevant attributes is termed as the curse of dimensionality. This motivated us to propose a similarity metric which is based on distinct weights and values based on commonly rated items.

The primary function of the newly introduced similarity measure is to increase the precision of established prediction systems. A collection of experiments on the various sizes of data sets of a different grade of sparsity have been performed to justify the accuracy and efficiency of the metric. The blueprint of the proposed similarity metric based on weight and value is discussed via Equation 6.2 and the selection of most significant neighbours for predicting the ratings of unknown items is performed by using Equation 6.3.

$$\text{Sim}_{\alpha}(x, y) = \sum_{i=0}^{m-n} \frac{1}{m-n+1} \alpha^{i} . \beta^{i}_{x,y} \tag{6.2}$$

$$\text{Similarity Index } (SI)_x = \frac{|I_x \cap I_y|}{(\text{Max} - \text{corated items})_x} * \text{Sim}_{\alpha}(x, y) \tag{6.3}$$
$$\text{Top}_n(SNN) = \text{Top}_n(\forall (SI)_x)$$

In Equation 6.2, Simα(x, y) represents the similarity between user 'x' and user 'y' such that x,y ∈ U. U denotes a universal set of users that are present in the system. α and β are two main components representing the weights and values respectively. 'm' and 'n' are the maximum and minimum possible rating scores that can be given to any item in the system by any user. In Equation 6.3, Ix and Iy represents the items rated by user 'x' and user 'y' respectively. Topn(SNN) denotes the top 'n' significant nearest neighbours for any user present in the system.

The two principle components of proposed similarity metric includes:

1. Values (β): The users show their interest in any particular item by providing a specific rating score to the item. In most of the cases, the range of rating score varies from (+1 to +5) or (+1 to +10), where +1 denotes that the user is least interested in the item and high score values denotes that the user is highly interested in the item. 'NA' values in the dataset represents that the item hasn't received any rating from the user.
   Suppose in a system we have 5 users and 10 items, The vectors representing the rating behaviour of user X_1, X_2, ..X_5 is given by:
   R_1 = {2, 1, NA, 3, 5, 1, NA, 2, 3, NA}
   R_2 = {3, 3, 2, 3, 4, NA, 1, 3, NA, 4}
   R_3 = {4, 2, NA, 4, 3, 1, 2, NA, 2, 3}
   R_4 = {2, 3, 1, 5, NA, NA, 5, 3, NA, 3}
   R_5 = {5, 4, 2, 2, NA, NA, NA, 2, 3, NA}

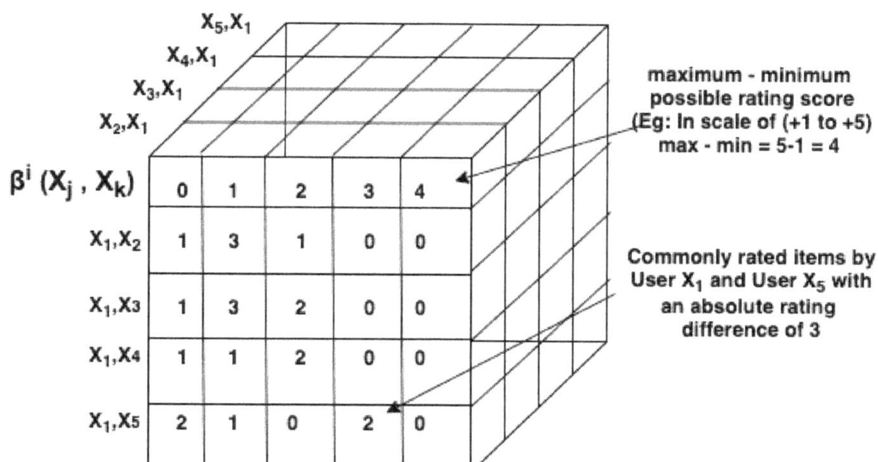

Figure 6.1 β vector values for commonly rated items

Figure 6.1 illustrates the obtained value vector by analysing the commonly rated items by user X_1, X_2, X_3, X_4, and X_5. The dimension of β is (User x R_T), where User denotes the total number of users (rows) and RT denotes the possible ratings an item can obtain (column). For e.g.: In above case, the rating values lie in the range of +1 to +5. Therefore, RT value is equal to 5. The mathematical representation for β can be computed using Equation 6.4.

$$\beta_{X_j,X_k} = \{\beta^0_{X_j,X_k}, \beta^1_{X_j,X_k}, \beta^2_{X_j,X_k}, \beta^3_{X_j,X_k}, \ldots \beta^{m-n}_{X_j,X_k}\} \quad (6.4)$$

where, X_j, $X_k \in U$ (Universal set of users present in the system), m and n represent the maximum and minimum possible rating that can be given to any item in the system respectively. The resultant value of β is obtained by dividing it with total co-rated items. Mathematically it is represented via Equation 6.5.

$$\beta^i_{X_j,X_k} = \frac{(X_j, X_k)^i_{A.D.}}{(X_j, X_k)_{C.R.}}, \forall i \in [0, 4] \quad (6.5)$$

where, $(X_j, X_k)^i_{A.D}$ denotes the absolute difference rating 'i' given by the two users X_j and X_k. $(X_j, X_k)_{C.R}$ represents the total number of items that has been rated by both user X_j and X_k.

For e.g.: In the above discussed rating scenario, the value of β for user X_1 and X_2 will be equal to (1/5, 3/5, 1/5, 0, 0). The value '5' in the denominator represents that the total number of items which has been rated by both user X_1 and X_2 is 5. The numerator represents the number of items that has received the same absolute difference ratings. The first value of β i.e., 1 denotes that one item has been rated by both user X_1 and X_2 with an absolute difference of '0' or in other words, '1' item has been rated identically by both user X_1 and X_2. Similarly, the second value of β i.e. 3/5 represents that 3 items has been rated by both user X_1 and X_2 with an absolute difference of '1'. Likewise other β values has been calculated by the system.

2. Feature Weight (α): The weight parameter of our proposed similarity metric helps in emphasizing the importance of β values which are obtained in the previous step. Each β value is associated with an α value that represents its level of importance while calculating the similarities among the users. In our proposed model, α ∈ [-1,+1], such that -1 and +1 represents high negativeness and high positiveness respectively. Figure 6.2 illustrates the efficiency of different α values in the solution space. In order to decide the optimal feature weights, we have used several optimizations (meta-heuristic as well as gradient based) techniques. Detailed description of the optimization techniques used by the proposed model is discussed in Section III.

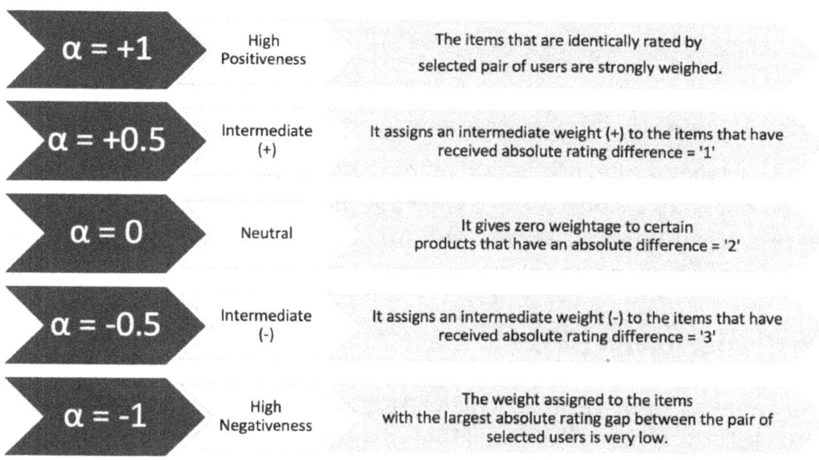

Figure 6.2 α values for commonly rated items

6.3 Optimization Techniques

We conducted a variety of tests utilising multiple optimization methods in order to adjust the hyperparameters of the framework. In the forthcoming subsections, the functionality and role of different optimization approaches are discussed. For deciding the optimal feature weights (α), we have used Mean Absolute Error (MAE) as the fitness function in our model. Equation 6.6 demonstrates the mathematical representation of the fitness function.

$$\text{Fitness Function}(f) = \frac{1}{N_X} \sum_{X \in U} \frac{\sum_{i \in I} |P_r - A_r|_i}{N_I} \tag{6.6}$$

where, N_X and N_I denotes the total number of test users and test items present in the system respectively. 'U' represents the universal set of users (test). The predicted rating value and the actual rating value is denoted by P_r and A_r respectively. The rating for unknown items is predicted using Equation 6.7.

$$P_{x,i} = \bar{R_x} + \frac{\sum_{y \in KS_x} [Sim_\alpha(x, y) * (R_{y,i} - \bar{R_y})]}{\sum_{y \in KS_x} Sim_\alpha(x, y)} \tag{6.7}$$

where, $\overline{R}_x$ and $\overline{R}_y$ denotes the average rating given by user 'x' and user 'y' respectively. KS_x is the set of users which are very much similar to user 'x'. In our model we have used the phrase "Significant Nearest Neighbours" to refer the top similar users. Simα(x, y) is the similarity between user 'x' and user 'y'.

Genetic Algorithm (GA)

The genetic algorithm (GA)[GK20] has been used extensively in many RSs as a heuristic search algorithm that imitates the mechanism for natural evolution. It is an iterative analysis approach focused on the analogy of natural selection and the biology of evolution. The existing population is substituted by a sequence of offsprings that have extremely appropriate hypotheses for any instance of the genetic algorithm. The fitness function checks the consistency of the hypotheses at each iteration. In the GA optimization process, a set of hypotheses also recognized as population is tested instead of only one hypothesis. The values of hypotheses are viewed as binary strings that are compatible with normal evolutionary genetic chromosomes. GA basically comprise of five major components namely: Population, Estimation, Reproduction, Crossover and Mutation. The GA population represents a collection of gene chromosomes (a set of values). The chromosomes possessing high fitness value are chosen to be kept in mating pool by imitating the natural selection. The reproduction process is responsible for copying the chromosomes to the next generation. Primarily focused on the combination process, the crossover operation produces children from the parents and the mutation is structured to preserve genetic variation from one generation to the next. A single GA generation involves one genetic operator loop iteration and fitness value estimation. The working principle of genetic algorithm is illustrated by Figure 6.3.

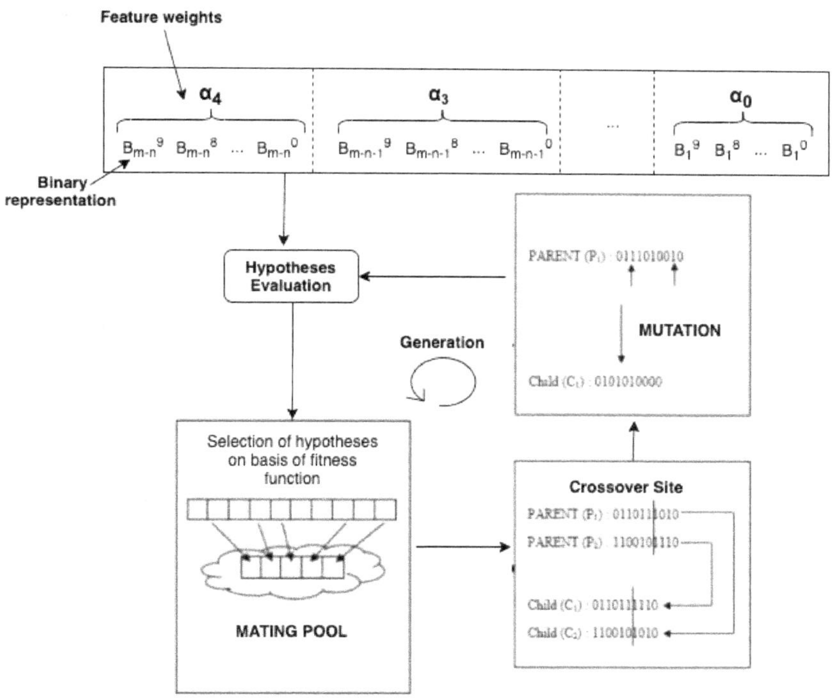

Figure 6.3 Genetic Algorithm

1. Reproduction: "Survival of the fittest". This process doesn't create any new string. It evaluates the fitness value of different strings and them to the mating pool.
2. Crossover: Crossover probability (P_c) tends to determine the preference of a parent from the match pool for crossover between the parent strings.

3. Mutation: This genetic operator aims to guarantee no essential genetic knowledge is missed in a timely manner. It also plays a significant role in sustaining population diversity. It decides whether to mutate any bit or not with mutation probability (P_m).

In order to convert the bitwise representation of feature weights back into the decimal form feature weights, we have used Equation 6.8.

$$\alpha^i = \frac{2\sum_{j=0}^{9} 2^j * B_i^j}{2^{10} - 1} - 1; i \in [0, 1..4] \tag{6.8}$$

Particle Swarm Optimization (PSO)

PSO[JD19] is also a population-based optimization technique like the genetic algorithm. As a substitute of using crossover and mutation to produce new offspring, the velocity and acceleration in PSO modifies the location of the particle to search the space of all possible solutions. In the region where the most suitable solutions are, particles can inevitably swarm around the region due to the rules controlling the swarming process. The initial location and velocity values of each particle are chosen randomly. Further these values are updated in the algorithm using two best values namely 'P_{best}' and 'G_{best}'. P_{best} denotes the best solution achieved by any particle at any instance. G_{best} represents the best solution obtained overall with respect to all the particles present in the population. The update on velocity and position of particles is done by using Equation 6.9.

$$V_i^{j+1} = \omega.V_i^j + K_1 * R * (P_{best}^i - Z_i^j) + K_2 * R * (G_{best} - Z_i^j)$$
$$Z_i^{j+1} = Z_i^j + V_i^{j+1} \tag{6.9}$$

where, V_i - Velocity of the particle 'i'
Z_i - Position of the particle 'i'
K_1 and K_2 - Constants
P^i_{best} - Personal best solution of particle 'i'
G_{best} - Global best of particle 'i'
ω - Inertia factor j – Iteration

Crow Search Algorithm (CSA)

A meta-heuristic analyzer, abstracted from the canny action of crows, is the Crow Search Algorithm. CSA[Ask16] is a population-based solution that is put together with respect to the root concept of how crows store their surplus of food in their shelters and retrieve a substantial amount of food when needed. Crows are regarded as one of the most intelligent birds. As compared to their body size, their brain is comparatively larger. Crows were observed to follow other birds, to examine where the other birds store their food, and to eat it after the owner leaves. If the Crow has performed a thievery, special measures would be made, such as changing hidden areas to prevent becoming a future victim. In reality, they use their own history of being a thief to anticipate the actions of a pilfer, and they may decide the best path to defend their resources from being looted. The working principle of Crow Search Algorithm is as follows:

1. Crows form a flock and retain the location of their hidden food.
2. Crows follow one another to perform wrongdoing (thievery).
3. Crows protect their food storage location from thievery by certain probability.

By following the aforementioned principle of CSA, following pseudo code is used to predict the optimal solution present in the search space [SD20].

CROW SEARCH ALGORITHM (CSA)

Initialization
Flock size : N
Position of Flock : - Random
Memory of crows : - Random
Gliding length : - G_l
Probability of awareness : - AP_j; where, j represents cr
Maximum Iteration :- I_{max}
Solution estimation : While $I < I_{max}$
    $\forall$ c $\in$ [1:N], c: crow
        choose crow 'j' : Randomly (to be followe
        $Z_{c,I+1} = R_c * Gl_{c,I} * (M_{c,I} - Z_{c,I}) + Z_c$
        $Z_{c,I+1}$ = Random position; otherwise
    Feasibility test
    Fitness Evaluation
    Memory updation
        $m_{c,I+1} = Z_{c,I+1}$; $f(Z_{c,I+1}) > f(m_{c,I})$
        $m_{c,I+1} = Z_{c,I}$; otherwise

The movement of crows in solution space can be represented by Figure 6.4[JSS19]. CSA possess a demerit of premature convergence especially in case of high dimensional datasets. This ultimately leads to inaccurate predictions and thus affects the accuracy of the model to a large extent. This motivated us to propose a new improvised crow search algorithm that uses Levy flight, accustom of commutable operator and proceedings component to enhance the exploitation and exploration of CSA.

Improved Crow Search Algorithm (iCSA)

The performance of CSA is improvised by introducing three vital components namely: Levy flight [Lf(n)]], accustom of commutable operator[φ] and proceedings component[ρ].

Ly(n): Levy flight promotes diversification in the solution space. Equation 6.10 denotes the mathematical representation for computing the diffusion value of levy flight.

$$\mathrm{Lf}(n) = 0.01 * \frac{R_1 * \left(\frac{\gamma(1+\beta) * \mathrm{Sin}\left(\frac{\pi * \beta}{2}\right)}{\gamma\left(\frac{1+\beta}{2}\right) * \beta * 2^{\left(\frac{\beta - 1}{2}\right)}} \right)^{\frac{1}{\beta}}}{\|R_2\|^{\frac{1}{\beta}}} \tag{6.10}$$

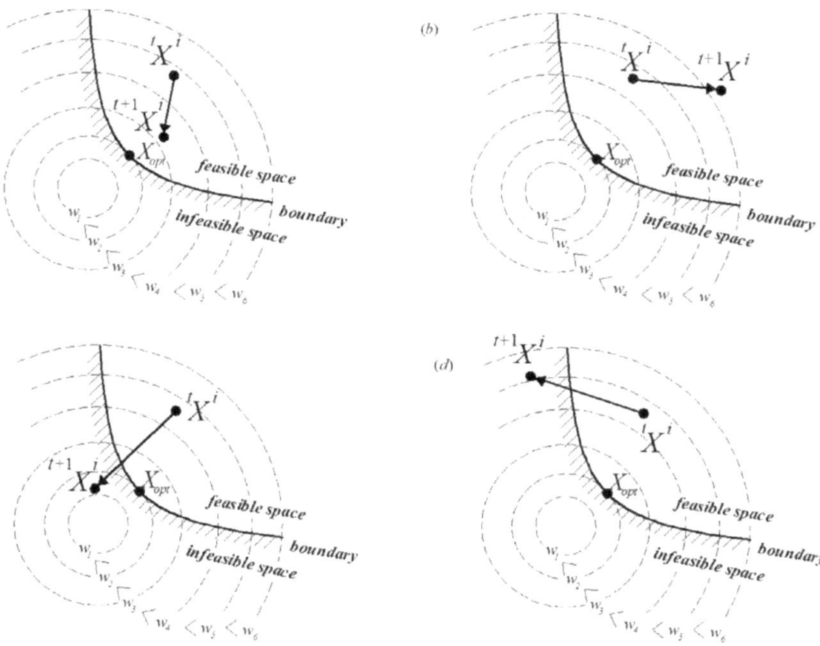

Figure 6.4 Possible crow motion in each iteration 'I'

where, R_1 & R_2 : Random numbers [0,1], β : constant and γ : gamma function.

The convergence efficiency is increased by the accustom of commutable operator and proceeding component. 'ρ' represents the experience of crows to discourage thievery of their food by using their own experience as a thief. In order to estimate the solution and update the position of crows at each iteration of the algorithm, Equation 6.11 is used.

$$\begin{aligned} Z_{c,I+1} &= R_c * Gl_{c,I} * (M_{j,I} - Zc, I) + \rho * Z_{c,I}; [R_j > AP_{j,I}] \\ Z_{c,I+1} &= \rho * Z_{c,I} + Lf(n) + \phi(1 - \frac{c}{c_m}); otherwise \end{aligned} \quad (6.11)$$

Gradient Based optimizer (GBO)

The GBO[ABC20] uses two key components namely: Local Escaping Operator (LEO) and Gradient Search Rule (GSR) to explore the solution space. The GSR uses the gradient-based approach to strengthen the exploration pattern and speed up the convergence rate to obtain better search space positions. LEO helps the GBO in escaping from getting stuck at local optima.

Gradient Search Rule

The movement of vectors is regulated in the GSR to obtain better positions in the feasible domain. The GSR is based on the grounds of the definition of the GB approach with a view to strengthening the discovery pattern and accelerating the GBO convergence. This rule is however, derived from the Newton method dependent on gradients. A computational gradient method is a replacement for a direct functional derivation since certain optimization problems really aren't differentiable. The GBO method typically starts with a hypothesised initial solution and travels in a gradient-specified trajectory to the next location. The mathematical formulation for GSA is represented by Equation 6.12.

$$GSR = R * \rho * \frac{2 * \delta Z * Z_c}{Z_{worst} - Z_{best} + \epsilon} \quad (6.12)$$

where, R:- Random number
$c \in [0,0.1]$
ρ:- Random parameter

Z_{worst}:- Worst solution
Z_{best}:- Best solution
The random parameter ρ is added to GBO to improve the search capability, local exploration and global exploration.

Local Escaping Operator (LEO)

The introduction of LEO in GBO helps in solving the complex problems easily. It generates new solutions by using best solution (Z_{best}) and two random solutions (Z_1^c & Z_2^c) present in the solution space. This LEO helps GBO in possessing a random behaviour while selecting the parameters and thus directs the population towards increased diversity which saves the algorithm from falling into trap of local optima [ABC20].

6.4 Illustration of Proposed Approach

In this section, we will discuss the nine phases of our proposed approach. The flow diagram of the proposed approach is demonstrated via Figure 6.5. The proposed model starts with data preprocessing/transformation phase and ends up with rating prediction of unrated items. We have used several meta- heuristic and gradient based optimization techniques to decide the optimal parameter values of the proposed similarity metric. The efficacy of the proposed model is reflected via low prediction error obtained in the last phase. The functioning of each phase along with the input and output is discussed as follows. For testing the performance of our model, we have used different datasets having different sizes. Table 6.2 shows the list of datasets which has been used by us to test our proposed model. In the first phase of the proposed model, we convert the extracted user-item information from the dataset into an appropriate utility (user-item rating) matrix. We replaced the unknown rating values by 'NA (Not Available)'. The rows of matrix represent the items and the column of utility matrix represents the users present in the system. Once the transformation phase is completed, the second phase of proposed model deals with splitting of dataset into 70:30 ratio, where 70% and 30% denotes the training and test set respectively. It is observed from Table 6.2 that most of the rating datasets are highly sparse. It is very difficult to train the model in such case. Therefore, in third phase, we used matching imputation technique to reduce the sparsity of the training set. The mathematical representation for imputing the rating values is given by Equation 6.13.

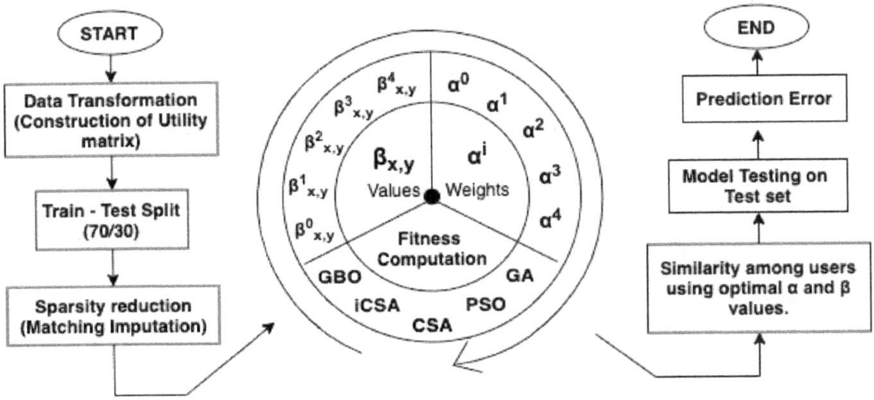

Figure 6.5 Flowchart of proposed approach

Table 6.2 Dataset Description

Dataset	#Users	#Items	#Ratings	Scale	Sparsity (%)
MovieLens (100K)	943	1682	100,000	+1 to +5	93.70
MovieLens (1M)	6040	3900	1,000,209	+1 to +5	95.8
MovieLens (10M)	71,567	10,681	1,00,00,054	+1 to +5	98.7
MovieLens (20M)	1,38,493	27,278	2,00,00,263	+1 to +5	99.47
jester-data-1	24,983	100	1810455	-10.00 to +10.00	27.53
jester-data-2	23,500	100	1708993	-10.00 to +10.00	27.27

$$(x_u)_r = \sum_{s=1}^{S} (x_u)_s; S \in (\text{Set of top} - \text{similar items w.r.t } 'r') \quad (6.13)$$

$$r\&s \in \text{Training set}$$

where, U represents the universal set of users and items present in the system. $(x_u)_r$ and $(x_u)_s$ denotes the rating given to item 'r' and item 's' by selected user 'u' respectively. The fourth, fifth and sixth phase of proposed model deals with estimating the optimal parameters of the new similarity metric using different meta-heuristic and gradient based approaches. Once we obtain the optimal weights and values to be used by new similarity metric, we calculate the similarity among different users present in the training set at the eighth phase of the model. The last two phases help in testing and analysing the proposed model performance. MAE (Mean Absolute Error) is used in our model to measure the error in rating prediction of unrated items.

6.5 Experimental Outcomes

We have performed the implementation of proposed model using R language. The model efficacy has been tested on MovieLens 100K, MovieLens 1M, MovieLens 10M, MovieLens 20M and Jester dataset. Hereby, we have investigated the outcomes obtained after completion of each phase of the proposed model. The constructed utility matrix and train-test split can be represented via Figure 6.6. After splitting the dataset into training and test set, we performed sparsity reduction of training set by using matching imputation technique. The reduction in sparsity level for different datasets is discussed via Table 6.3. We didn't performed imputation on Jester dataset as the dataset is having less records and unnecessary imputation will lead to over-fit the dataset.

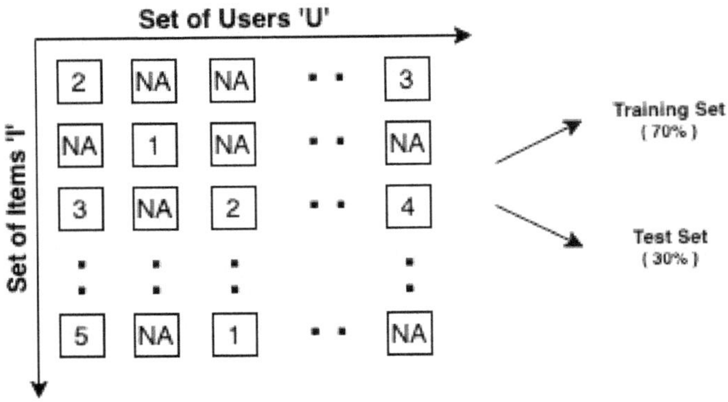

Figure 6.6 Utility (USER-ITEM rating) matrix

After reducing the sparsity in the training set, we computed the value of β for all possible pair of users present in the system. The method of calculating the β value has already been discussed in Section II. The α values lies in the range of [-1,1], where -1 denotes negative weight and +1 denotes the high positive weight.

Table 6.3 Sparsity reduction using matching Imputation

Dataset	**Reduction in Sparsity (%)**
MovieLens 100K	39.30
MovieLens 1M	42.30
MovieLens 10M	43.77
MovieLens 20M	47.21

Experimental setup for Optimizing 'α'

Parameter details for all meta-heuristic approaches that has been used for optimizing α are as follows:
Population size = 100
Maximum iterations for convergence = 100
Seeded region of α : α0[+0.6, +1], α1 [+0.2, +0.6], α2[-0.2, +0.2], α3[-0.2, -0.6], α4[0.6, -1.0].
Crossover Probability (P_c) = 0.8 Mutation Probability (P_m) = 0.2 Inertia factor (ω) = 0.1

Convergence of different optimization techniques

The convergence of different optimization technique for obtaining the optimal α value can be depicted in Figure 6.7. It is clear that all the optimization technique which has been used in our model to estimate the α values converge at around 100 iterations. Therefore, in order to show the convergence, we iterate the model for 150 iterations. The α value obtained after 100 iterations is used further to predict the rating of unknown items from the test set. In our model we have used 100 iterations because all models converge around 100 iterations and it reduces the time complexity of finding the optimal solution.

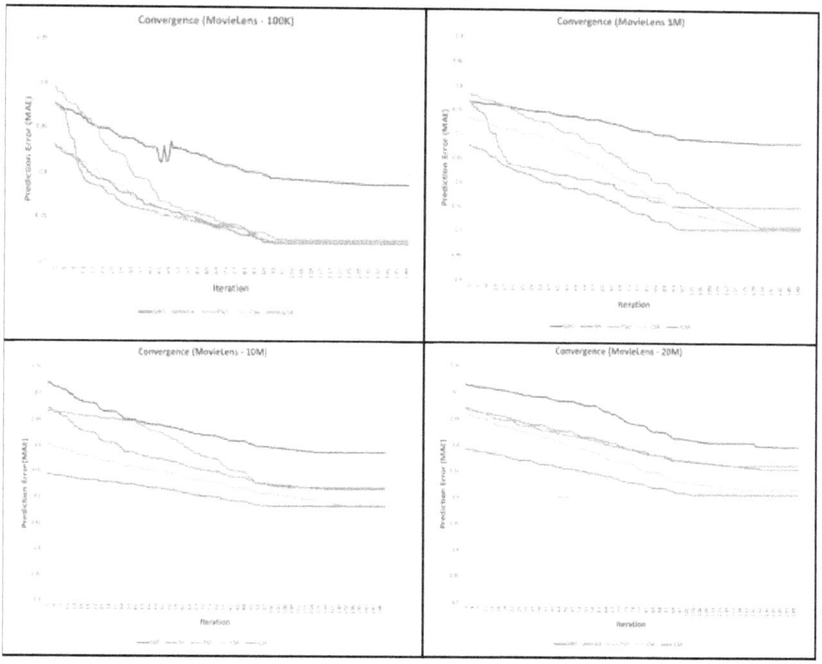

Figure 6.7 Convergence Graph

Comparative analysis of different optimization techniques

We have compared the outcomes of proposed similarity metric with two most widely used similarity metric i.e., Cosine similarity and

Pearson Correlation Coefficient (PCC). We have summarized the performance of different optimizers on different dimension of datasets in Table 6.4. It is observed that the proposed similarity metric outperforms the traditional similarity metrics in predicting the unrated movies. We also observed that the iCSA provided slightly better results as compared to other optimization techniques. There is no negative impact of high dimensional datasets in case of the proposed model. The proposed model works absolutely well even when we are working with 20M MovieLens datasets. The introduction of Levy flight, accustom of commutable operator and proceedings component helped in better convergence of iCSA towards the optimal solution as compared to the traditional crow search algorithm.

The iCSA algorithm is compared not only according to the solution values obtained, but also according to the distribution in the search space of the solutions obtained. In addition, the information it presents will help define the exploitation and exploration capabilities of the algorithms that are compared. Deep statistical comparison (DSC) [EKS17] and Extended deep statistical comparison (eDSC) [EK19] is used for comparing the meta-heuristic stochastic optimization algorithms. In order to regulate family-wise error (FWER), Bonferroni correction [Gar+09] is used. Table 6.5 demonstrates the statistical comparison using eDSC ranking scheme with respect to solution values and distribution of solutions in search space. The $P_{value-Y}$ & $P_{value-X}$ correlates with P_{value} for comparing the solutions obtained and solution distributions in search space respectively. Smaller value of $P_{value-X}$ and $P_{value-Y}$ reflects that the iCSA is statistically significant as compared to other stochastic optimization techniques.

Table 6.4 Comparative analysis of similarity techniques

Similarity	**Jester-1**	**Jester-2**	**ML-100K**	**ML-1M**	**ML-10M**	**ML-20M**
Cosine	1.0985	0.79280	0.75240	0.59579	0.57431	0.61138
PCC	1.0713	0.76650	0.75491	0.59556	0.56829	0.58745
Sim: GBO	0.98823	0.76013	0.79321	0.68972	0.59713	0.61246
Sim: GA	0.97620	0.75991	0.72312	0.55085	0.52476	0.57126
Sim: PSO	0.97170	0.75720	0.72488	0.57826	0.52132	0.56348
Sim: CSA	0.81327	0.67392	0.72251	0.53661	0.50648	0.52793
Sim: iCSA	0.80712	0.67271	0.72101	0.50374	0.48321	0.50832

Table 6.5 Statistical Comparison of iCSA vs GBO, GA, PSO, and CSA

Algorithms	Pvalue−Y	Pvalue−X
GBO, GA, PSO, CSA	0.033	0.087

The context of recommendation is decided by checking the timestamp (in case of MovieLens dataset) and we search for the item that is most recently viewed or rated by the user. Then we check the item which is most similar to the item in which the user has recently showed interest and accordingly we generate the list of items to be recommended to the user.

6.6 Results & Discussion

By comparing Mean Absolute Error value with a number of state-of-the-art comparative recommendation models, we display our findings through Table 6.6. The implemented meta-heuristic approach has taken the size of population as 100. The results obtained from each optimizer indicates that the performance of proposed similarity metric with iCSA is better than other traditional similarity metrics. GA performed better in making predictions for jester dataset which has relatively low sparsity level as compared to MovieLens datasets. But practically, in real world scenarios, we will come across the datasets which are highly sparse because most of the users don't rate most of the items present in the system. In such situation, the efficacy of GA will degrade to a great extent. iCSA plays a key role in providing rescue for such highly sparse datasets. The error value obtained by our proposed model is 0.80712, 0. 67271, 0.72101, 0.50374, 0.48321, and 0.50832 for jester-data-1, jester-data-2, MovieLens-100k, MovieLens-1M, MovieLens-10M, and MovieLens-20M respectively.

Table 6.6 Comparison of proposed model with state-of-art models

Dataset	Recommendation Technique	MAE
ML-100K	Auto-Surprise (TPE)[AB20]	0.7280
ML-100K	Auto-Surprise (ATPE)[AB20]	0.7244
ML-	FCM with PSO & Kmeans [KV17]	0.7547

100K		
ML-100K	PCA-GAKM[KV17]	0.7821
ML-100K	**Proposed model**	**0.72101**
ML-1M	BiUCF[KM18]	0.7379
ML-1M	PMMF[Kum+17]	0.6760
ML-1M	DsRec[Liu+15a]	0.6790
ML-1M	**Proposed model**	**0.50374**
ML-10M	DLSCF-S[JDF19]	0.612
ML-10M	RSVD[Kor08]	0.6438
ML-10M	**Proposed model**	**0.48321**
ML-20M	DLSCF-S[JDF19]	0.6260
ML-20M	GIDS-RMF[JCB18]	0.66
ML-20M	**Proposed model**	**0.50832**

6.7 Conclusion

Similarity metrics were one of the main factors in improving the prediction accuracy. In this article, five variants of optimization methods have been applied focused on naturally motivated approaches and gradient based approaches. In order to address premature convergence constraints and inability to achieve global optimisation, we have strengthened CSA's approach to high-dimensional dataset. The method proposed combines similarity metrics with optimised hyperparameter using iCSA to produce a highly efficient user profile. The efficiency of the model was analysed and contrasted to a variety of comprehensive strategies for personalization. The comparative findings reveal that iCSA provides the best results in terms of MAE value. In future, for further insights, deep learning can be used to analyse the changing behaviour of users.

CHAPTER 7

Amalgamation of Social and Trust-Aware Recommender System with Deep Learning

"Social media is not just a spoke on the wheel of marketing. It's becoming the way entire bicycles are built" — Ryan Lilly

The strength of social networks lies in their potential to link people in the most convenient manner and then suggesting the relevant information to them. Although by developing these types of functionalities, we at the same time are providing support for other aspects, including the identification of similarities and ability to harness the information of social relationships based on the connections and the network's infrastructure and the number of links. The term social background includes different forms of side detail, such as social tags or internet trends. Often, unique context-sensitive RS may be used as a particular form of a more general-purpose type of recommender systems. The social context incites the presence of several human influences, such as trust crucial function in having the consumers recognise the actors participating throughout the feedback process. Social networks typically employ different methods when providing you with new people, classes, websites, or contacts. This would be via "PYMK" and "mutual friends" in Facebook, LinkedIn, Google+, etc.

7.1 Introduction

With the emergence of Web 2.0, social RS have emerged as a powerful field of research. Traditional RS approaches do not usually take into account the social experiences, which make it more difficult for them to further improvise. A crucial point that bears emphasis that is essential in this statement is that when we are confronted with several options, we would seek out our other friends that we can seek out immediate guidance from. More precise details should be integrated in order to maximize performance; thus, the social network information should be accounted for. Based on the aforementioned points, few researchers [JE10] [OP13] proposed few trust-based RS. These methods make use of trust information to further improvise their performance. But these approaches have their own issues, which need to be addressed. There is one apparent shortcoming that stands out: the "trust relationship" issue is broken. The notion of "social ties," which relates to the cooperation and mutual interaction between users, is different. In comparison, the other shortcomings are the utterly impossible theory and the poor generalization capability. Obviously, trust-based advising programs may no longer be relevant. The analysis of actual social-based advice networks is then seen on the display. In addition, the integration of social networks will potentially increase the efficiency of existing advising programs. First, in terms of predictability precision, friendships between users boost the comprehension of user scores. We may then view consumer expectations more accurately. Second, as a matter of truth, relationship between the two users suggests that they will have interests in common. As a result, the cold-start issue may be mitigated.

In order to solve the problem discussed above, [Sun+15] proposed RSboSN that combined social-based graphs and utility(user-item) matrix to increase the precision in predictions of conventional RSs. In the recommendation procedure, they primarily use the friendships between users and the user-labelled tags to generate recommendation. The user-item-tag can be viewed as 2D matrix. They clustered similar users to measure the similarities between users and to calculate correlation between user and item. The purpose of clustering is to find the most appropriate friends for realistic recommendations. On the basis of the method discussed by a researcher in [Ma+11], the two detailed facets of social network knowledge referred to above are included in the design of the terms of social regularization.

Social knowledge contains, on one hand, ample consumer attributes and interests, such as reviews, feedback, communities and user associations, which can be utilized to enhance recommendations; on the other hand, the recommending mechanism can significantly assist the buying and booming of products and customers and keep the social network vibrant. As a result, how to take account of new contextual knowledge to develop commendation frameworks seems to have become a great attraction in the corporate and scientific fields. There are many more consistency of choice amongst friends of common interests, which seems to be the underlying motivation of the social recommendation system. Social recommendation systems that create social networks via social contextual knowledge between users may enhance CF and resolve problems of sparsity and cold start in recommender system. Figure 7.1 displays all the social qualitative knowledge that can be obtained from social network connections. For new users, a recommendation may be made through social connection and established users as long as long as there is some relationship between the new and existing users. A proven and tested strategy exists on one side, while scientific data reveals that inter-linked hobbies and behavioural standard tendencies exist in online communities, and while interrelationships demonstrate another. Furthermore, through the observational findings of the social network [DGM08], the theoretical social advice has proved accurate and right, which definitely strengthens the social contextual knowledge framework in its recommendations.

In recent years there have been tremendous changes on the media platforms, the interactions and association between users are increasingly multiplying. For instance: Let W, X, Y and Z are four users present in any system. The interaction details among them are as follows:

W follows X & Y
X trusts Y

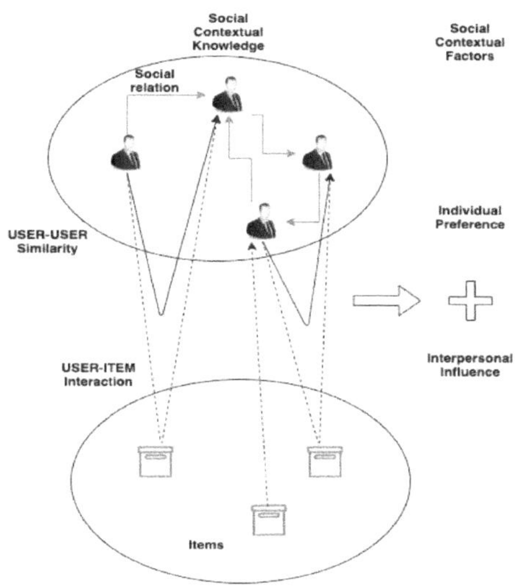

Figure 7.1 Social relationship among users and items

X and Z belongs to same group 'G'
Y always comments on Z
W approximates Z

The aforementioned dynamic behaviour of users is illustrated via Figure 7.2. The perplexing combination of social networks also brought new problems and opportunities to light which are as follows:

1. Diversity is seen in social relations much more often. there are various kinds of relationships existing among users of social media, such as friendly relationship, interpersonal trust, co-group, etc. Many distinctive influen- cers exercise their influence, and exert user behaviour simultaneously. It's a wonderful chance to aid in enhancing recommendation performance by integrating various social networks.
2. Fusing various social networks is very complex, which makes it non- linear and impossible to evaluate. Social effect on consumer behaviour is nonlinear and is distinct. The idea of how to develop recommendation based on various kinds of social networks is a challenging task.
3. The implicit social network may extract users' temporal and spatial knowledge. We may take the features of the users' spatial and temporal position knowledge and convey it as an implied social network. For instance, the similarity of users can be interpreted as a graph where each user is a node, and the association of users' positions is a social network in which edges reflect their proximity to other users' frequency of interactions. A similar pattern can be observed in the network that shows how users communicate behaviourally, and locations can be observed as a manner in which they relate to each other.

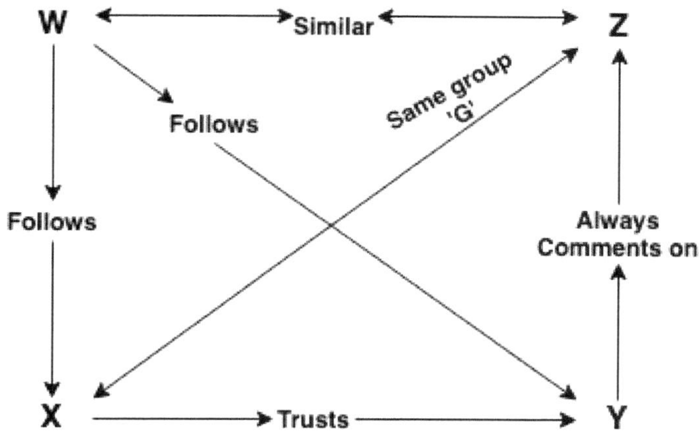

Figure 7.2 User's complex relationship

Social recommendations have made significant advances in both the academic and industrial fields in the past decades. As more and more commercial entities adopt social platforms, sociological research on the recommendation method is making great headway, meanwhile, the infrastructure of the recommendation system has seen significant growth, along with it have been advancing scientific concepts of decision support. However, the nuances and varieties of social suggestion are seldom anastomosed with a single social network informally. Enhancing the recommendation from various social networks has considerable importance. Today, diverse research suggests on various social networks [Wu+16a] [ZQX16], while a linear approach disregards social meaning importance and interconnection. Targeting at these challenges, [YL18] proposed a framework based on matrix factorization and joint probability distribution to various channels of social network.

7.2 Trust-based Recommender System

There is a significant degree of maturity in the mainstream approaches, as well as to their growth, but it's all predicated on users having full autonomy. Unfortunately, though, there are few true friendships in the consumer population. Based on these observations, many experts have sought to place more faith in social information and personal information guiding together to enhance the efficiency of conventional approaches. A confidence and credibility scheme ensures consumer satisfaction by offering details about how trustworthy the seller or buyer is. It establishes relationships between sellers and customers by using customer reviews and confidence scores. Future potential deals rely heavily on the obtained trust values. Following are the different challenges faced by trust-based systems:

1. Playbook Threat: Once the credibility of any particular user is boosted, he/she might misuse this high credibility as an opportunity to provide bad quality services to other consumers.
2. Unfair Ratings: This type of rating is provided by malicious attackers to mislead the decision-making process. These attackers deliberately provide unreliable ratings to the item to create an impact on other users.
3. Discrimination Attack: Discrimination usually happens in case when any user favours one group while providing rating scores, whereas gives low rating to other groups.
4. Re-entry Attack: The person with a low score could exit the group and return to escape the repercussions of being given a low reputation.
5. Sybil Attack: Any individual might create several copies of their identity, and making use of these different identities, he/she might provide a biased rating to any service. This will unnecessarily enhance the potential of single user effect on the system.

Using TRUST

Empirically, it is seen that people depend on input from trustworthy peers, and relatives, more than the automated recommendation method. The trust and commitment similarity possess positive relationship. Similarly, it is said that consumers like to get recommendations from others they know. This means that the usage of trust-based systems with other RS may improve recommendation efficiency.

The Advogato project is among the first experiments on trust-based advisory schemes. The central authority determines the individual's credibility in this endeavour. In this scheme people are marked as trustworthy and untrustworthy. By contrasting the former recommendation scores and the true scores, the trust score is determined indirectly. The previous user ratings is analysed to figure out exactly whether any particular user would be able to forecast the appropriate ratings of the products that can be recommended to other users.

Using DISTRUST

The term distrust can be used in three different ways:

1. Distrust can be used to screen out the inappropriate user from the system.
2. It can be used for debugging false positives (indicating the existence of a given condition, but actually it does not)
3. It acts as an indicator of negative trust.

7.3 Social-based Recommender System

The principle working of "social advisory systems" is described as the combination of knowledge about social networks that can impact online personal actions, like interactive information between users and tag information, in order to enhance the advisory systems. In recent years social network development has already been a buzz word and researched in several applications on how to use information available on social networks.

Let's assume a scenario where Jennifer wants to watch a TV series on weekend. Her favourite TV series genre is 'Comedy'. She finds two new TV series interesting on the internet namely: "The Crew", and "I Am Not Okay with This". Both these TV series are highly rated by IMDb. She is unable to decide which series to watch, so she calls her best friend Stacy with whom she often hangs out with. Stacy hasn't watched either of these two TV series but one of her colleagues had recently watched "The Crew" and was talking high of it. So, she recommends, "Why don't you watch 'The Crew' this weekend?". Jennifer follows Stacy's recommendation and enjoys 'The Crew' series on the weekend. After reviewing the aforementioned scenario, we conclude that the decision taken by Jennifer relies on three major factors:

1. Jennifer's own interest in comedy genre of TV series.
2. Review or public opinion on these two TV series.
3. Her friend stacy's opinion about both the series.

Recalling our common decisions in our day-to-day life, such as cafes, purchasing home, searching jobs, etc. is greatly affected by these three influencing factors. Figure 7.3 demonstrates the role of these three influential factors while deciding to purchase any product or use any service.

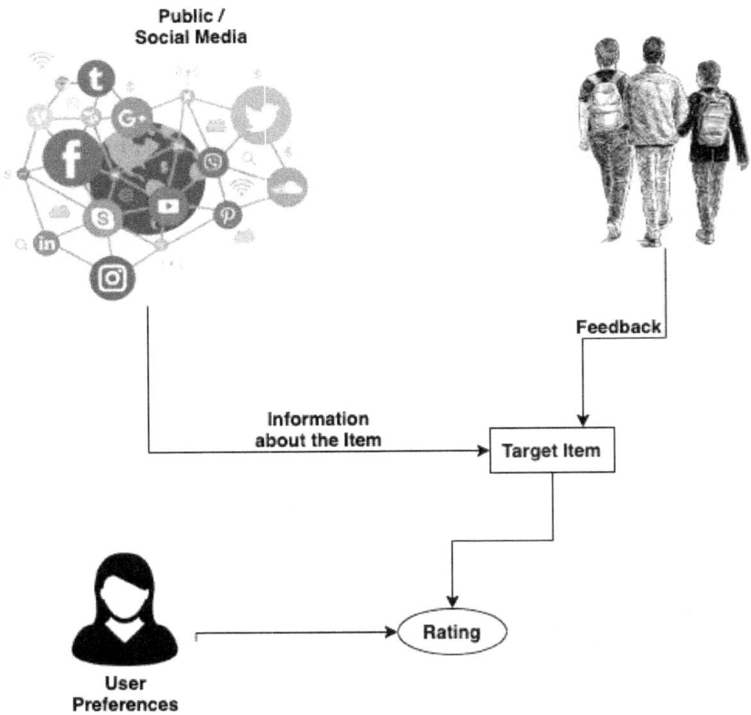

Figure 7.3 Influential factors in Social-based RS

7.4 Social Context from Trust and Network Centric Methods

The fundamental premise in network-centered approaches is that a user's amicable structure profoundly influences his tastes, preferences or consumption patterns. Users also ask for advice from their friends about film, tourism or other items. Moreover, social interactions show the common homophily, which also has identical preferences and tastes for connected users. This similarities in preferences also causes a consumer to trust their related users' suggestions rather than others. Several approaches can be developed to include certain links in the decision process. While such links might differ according to the problem domain, they are always very useful whenever a cold-start occurs, when the details on a user's ratings is minimal. In such scenarios, it can be particularly beneficial to recognise the related peers by information embedded in our social connections.

Data Collection

In the process of social recommendation, trust and homophily play a significant role. Both are having similar definitions, although not quite identical. Homophily refers to the idea that linked users are potentially identical in terms of preferences and desires on social networks. Trust ensures that people have a better likelihood of believing their mates' preferences and advice. Trust is perhaps the outcome of homophily in certain circumstances. Because linked users appear to be alike, they prefer to trust the preferences and advice of each other. To determine the trust between two user's 'X' and 'Y', we can use following similarity approach as discussed via Equation 7.1.

$$T_{X,Y} = \text{Cosine}(X, Y);\ if\ X\ and\ Y\ are\ linked$$

$$undefined;\ Otherwise \tag{7.1}$$

where, $T_{X,Y}$ determines the trust value associated between user 'X' and user 'Y'. Here we can observe that we are computing the trust values among the connected users only. Further in this chapter, we will see different trust propagation methods that can compute the trust scores of even the disconnected users present in the system. Equation 7.1 is inferred as an implicit way of computing the trust between users. In some scenarios, the trust might not be computing the implicit factors, in such cases the network asks the users to get involved in explicit trust links. For example: Epinions asks their users explicitly whether they trust or not the other users. Thus, by using this explicit information they build a trust network. On another website: Moleskiing, the users are explicitly asked to give feedback on how significant they found the reviews given by other users present in the system. This kind of feedback facility is also available on

Amazon.com. A trust network is built based on either implicit or explicit inferred trust relationships.

Trust recommender programs may render customized and reliable suggestions by using the information on the platform of trust. Such systems of recommendation are sometimes called trust enhanced RS. Many of these trust- enhanced RSs makes use of trust aggregation and propagation method to estimate the trust of unknown user in a transitive network. In simple words, if Jennifer(user-1) trust Stacy(user-2), and Stacy(user-2) trusts Pooja(user-3), how much trust will Jennifer(user-1) have on Pooja(user-3) can be estimated. Trust relations are asymmetric in nature, i.e., the level of trust Jennifer has on Stacy might not be equal to the level of trust Stacy has on Jennifer. However, in some cases where trust relationship is derived implicitly using Equation 7.1, the relation is symmetric in nature.

Trust Aggregation & Propagation

It is a matter of fact that most of the users on a social network doesn't necessarily have trust relationship between them. In such cases, transitivity property can be used to find out the trust relations between the users who have missing direct trust relations. The trust propagation is computed along a single path whereas, trust aggregation is computed across multiple paths. These methods act as unsupervised approach while generating recommendations, as they use predetermined heuristics in trust-centric systems.

1. TidalTrust: This type of algorithm is based on the ideology that shorter paths are more efficient and reliable for trust propagation. Thus, one should make use of shorter paths while computing trust relationship between a pair of source and sink. Figure 7.4 illustrates the working of TidalTrust algorithm for a trust-based network. In this figure, 'D' denotes the distance. The left side of figure represents the trust-centric network and the right side discusses the formation subgraph of shortest path between the source and sink present in the trust-centric network.

There are several algorithms that can be used in a trust-based recommender systems for generating the trust relations. They are as follows:

2. TrustWalker Algorithm: It is premised on the reality that social connections include knowledge that is distinct from ratings. Therefore, this algorithm uses random-walk to find out the similar set of users. However, a significant problem is that getting into a random walk can end up wasting valuable time on superfluous users.
3. MoleTrust Algorithm: Its working principle is similar to that of TidalTrust algorithm. The only difference is that in MoleTrust, there is no sink and the MoleTrust algorithm comprised of two forward passes as compared to one in case of TidalTrust algorithm. The termination criteria for computation of shortest path in this algorithm is with respect to path length.
4. Link Prediction: If you are just interested in predicting an ordered lists of preferred products, instead of predicting absolute values of individual item scores, you may use link methods. While most link forecasting methods may have strong rankings, they are not precise when it comes to identifying the weight of the links.
5. Matrix Factorization

7.5 Positive Aspects of Social Network

It's advantageous to integrate additional trust factors into social recommender frameworks. In particular, this is helpful for increasing the consistency of advice, fixing cold-start problems, and rendering the method immune to attacks. Some of the key merits of using social network in building a recommender system are as follows:

1. Improving the recommendation standard for controversial users/items is one of the best ways to increase the overall trust in the system. It may be controversial if several users have differing views regarding the ratings of the same thing. Highly controversial things are those which obtain critical feedback. Example: Similar users are more likely to trust each other and have high probability that both of them will provide similar ratings to controversial items as well.

2. For new users, social links serve as a lifeline, because they enable others to instantly connect with them. Suppose in a link prediction scenario, a new user enters the system and thus has no rating information regarding any items present in the system. In such case, no user-item relation is available with the system. While on other hand, if the user is associated with social ties, link prediction can predict the top items to be recommended. The crucial idea is that users' social networks are usually primed for use long before they begin to use the framework. This holds particularly true for implicit trust-based networks, as well as tacit networks. In any case, additional social ties contribute more data to deal with the recommendation sparsity problem.
3. More broadly, there are considerable economic benefits for suppliers to misuse recommender programs that are administered by 3rd parties. For instance, the producer of a product could attempt to give fake reviews for his product on Amazon. Often, such reviews are shared using bogus accounts of the manufacturer. Algorithms dependent on trust are less vulnerable to such attacks when the user-trusted recommenders are employed.

7.6 Social and Trust-Aware Systems with DL

In past few years, DNN have witnessed widespread success in several domains such as: NLP, Speech recognition, computer vision, etc. Several models based on deep learning has been proposed to enhance the efficiency of recommender system. GraphRec [Fan+19c] aggregated social interaction and user-item interaction using a graph neural network for predicting the interesting items for any particular user. Another researcher proposed DASO [Fan+19a] that hands over user's data between social and item domain using adversarial learning. The above listed deep social recommender systems benefit from social knowledge, but don't exploit it completely. These make only informal associations with close neighbours, although user data that is a hop or two away may still be useful. Following are the different reasons behind this:

1. Information is beginning to trickle via the social platform and we can impact users through non-targeted friends, followers, or indirect neighbours.

2. For consumers, remote connections (weak ties) may be used as intermedi- aries, but in turn the neighbors closely adjacent to them (or to relatives of intermediaries) lack knowledge to move on. Thus, remote social ties should be taken into consideration for recommender schemes.

The recommender method can use some but not all the input from neighbors is necessarily helpful when generating a specific recommendation. For instance, if any user is interested in buying a watch, then the interaction between his/her neighbours and watch might seem helpful in generating recommendation but the interaction between his/her neighbours and mobile phones will be of no help in enhancing the recommendation objective. Thus, filtering of information received from the neighbour seems to be an important task before considering it for the designed recommender system. Some of the challenges faced by existing social and trust-aware systems with DL are as follows:

1. The kind of relationships are difficult to contend due to the varied social interaction and, since you can't reliably determine if the guidance it provides would actually be useful for others or not.
2. Selecting useful knowledge from neighbours can be challenging, since there are many items with which they could have interacted with.
3. It is hard to know what the user cares about while you're estimating the user-item experiences.

DSCF [Fan+19b] addresses some of the aforementioned issues by exploiting the social information. This model presented outstanding performance and can be improvised further by using side information while analysing the user-item interactions.

7.7 Conclusion

In general, you can use social knowledge in any recommender schemes to extend the scope of implementation scenarios. The minimal expanded multidimensional model is useful such that it includes certain social factors, but only in a minor way. Other approaches that put trust at the heart of the framework can be used to build robust recommender systems. Trust aggregation and propaga-

tion techniques are used in an unsupervised setup. Using supervised methods yields better results, supervised methods such as relation prediction and MF are amongst the ones that possess high efficiency. By integrating the trust related information, the system becomes resilient to attacks, and also performs better for cold-start users.
Several researchers have proposed different deep learning-based models that makes use of social information. But still there is a long way to go in order to capture the complex relationship among the users present in the system efficiently. Few researchers suggested to use side information about the users and items apart from the rating scores. This side information acts as a jumping board for the recommender system while detecting the complex relations and ultimately helps in building an efficient trust-aware social recommender system.

CHAPTER 8

Recent Advancements and Application of Deep Recommender Systems

"Our future success is directly proportional to our ability to understand, adopt and integrate new technology into our work" — Sukant Ratnakar

This chapter highlights several advancements and application of recommender systems such as active learning, multi-criteria recommendations and privacy. Recommender models have developed several ways in the last few decades. In this chapter, we will cover some of the fundamental approaches employed in news personalization, advertising, query recommendation, etc. Many of the recommendations are carried out where multiple users or assessment requirements occur. One example of how a group of teenagers would like to watch one movie together, it's unlikely that anyone wants different recommendations, they might like recommendations that fit the entire group's interests. In the alternative scenario, users can offer ratings based on a variety of criteria. Depending on the problem formulation, these differences will make the estimation more difficult. Specifically, we will discuss the following advanced recommendation algorithms in this chapter: Ranking, Online learning with bandit problem, Multiple-criteria RS, Group RS, Active Learning, Privacy, etc. We have also discussed several application domains such as advertising, news recommendations, etc.

8.1 Introduction

There are numerous and diversifying (emerging and important) lines of RS study. The reader is strongly urged to turn to recent volumes of the ACM conferences and other superb review articles for further reading. Deep Learning played a key role in capturing the complex trend pattern present in the recommender system. While reviewing the literature, we found that the new deep learning approaches would produce an excellent basis for the recommender framework. Each DL model learns how to apply the given data in its own distinct way. Concisely, CNNs are effective for tasks like content-based feature learning, such as preventing features from being misclassified during initial classification from being too cold (cold-start problem). Because of its broad capabilities including multi-modal learning, prediction, time series forecasting, and sequence suggestion, the RNN is widely used in a variety of fields such as marketing and finance. Using the various forms of deep learning increases the conventional recommender systems overall efficiency. Finally, let's note that the deep learning recommendations are in the forefront of technological advancement, though that developments in this research trend are making new possibilities and challenges emerge. Some of the emerging trends in deep learning-based recommender systems are as follows:

1. Advanced Evaluation Approaches: It was discovered that a recommender scheme would not lead to a long-term improvement if it focuses only on increasing recall/precision or reducing prediction error. A new collection of evaluation metrics such as novelty, variety, coverage, serendipity, trustworthiness, and interpretability must be paid attention to. These set of assessments not only looks for achieving high accurate predictions but also aims at achieving high level of user satisfaction.
2. Scalability of the recommender framework: The scalability issue is concerned with the never-ending rise in data volumes of products and consumers. As a result, scalability is an important aspect when selecting a recommendation model. A recommender framework should be able to satisfy all conditions at the same time. Deep learning demonstrates its strength in big data analytics in this sense.
3. Amalgamation of different deep learning approaches: Several works focus on more than one form of neural network, with numerous deep learning strategies combined. In Chapter 2 of this book, we stated that not all possible hybrid approaches of deep learning have been explored. Many models based on more than one deep learning should be rigorously tested and with respect to their functional applicability. Some of the hybrid combinations of deep learning tested by several researches is demonstrated via Table 8.1.

Table 8.1 Combination of different DL techniques

Reference	AE	CNN	MLP	RNN
[Zha+17]		✓		✓
[Li+17]	✓			✓
[Zha+16]		✓	✓	
[RK16]	✓	✓		
[Lei+16]	✓	✓		
[WXY16]			✓	✓
[Lee+16]		✓		✓
[ESH15]		✓		✓

8.2 Advancements of Recommender System

Ranking

The different recommender models discussed in the preceding chapters dealt with recommendation estimation, in which the squared error is minimized to generate recommendation list for the user. But in general, only the top-N recommendation is provided to the consumers, and the other factors are dismissed. Thus, it's prudent to empirically refine ranking based criteria, including the mean reciprocal rating or the ROC curve. Besides that, the consumer places more emphasis to the first set of items on the list. The values obtained are inconsequential to the consumer. Optimizing the forecast values is often does not result in the right recommendation lists for the end user. This can happen, and therefore the resulting approach would not have a high-quality suggestion list to the end user. Furthermore, a prediction-based goal may provide limited solutions. This dilemma emerges because the forecast method objective functions don't meet the end user's needs.

The ranking schemes can be categorised into two measures:

1. Top-heavy ranking: Mean reciprocal rank, NDCG, etc.
2. Global ranking: Example: Spearman coefficient, AUC or ROC, etc.

One of the major challenges faced by ranking-based RS is its low capability for the non-smooth objective functions. It's very hard to optimize the system with gradient-based approach. A tiny adjustment in the predictions will result in dramatic variation in the rankings, and even objective functions. There is an adage which claims that the first rule of ranking is to estimate the rating with a loss function and then to go on to rank the items according to these estimated ratings. Following are the different methods for learning ranks:

1. Pairwise learning of ranks
2. Listwise learning of ranks

Multi-Arm Bandit

One of the key challenges faced by many real-time recommender systems is the continuous arrival of new items and users in the system. The recommendation approach needs to explore and exploit the search space on a real-time basis. Whenever an opportunity presents itself, the recommender system must distinguish between multiple options about whether to recommend to the customer. Depending on the domain, these options can be different. In our preceding chapters we discussed several offline based recommendation approaches where we are not much bothered about the new users or items entry into the system. The advancement which needs to be done is in the direction of reinforcement learning, in which both explore and exploit step is performed simultaneously. Multi-arm bandit algorithm is among one of the approaches that maintains the trade-off between exploration and exploitation. This algorithm mimics the gambler behaviour inside a casino where he tries to select a slot machine that might give him high payoff. The gambler cannot predict which of the machine can yield highest payoff as compared to others without playing on all the slot machines available to him/her. Several strategies that can be used by this algorithm to regulate the balance between exploration and exploitation includes: Naive approach, c- Greedy approach, Upper Bounding methods, etc.

Group Recommendation

This framework aims at making predictions of items in which an entire group might be interested rather than focusing at the consumption pattern of an individual. For example: In case of generating movie recommendation list, a group of users would like to watch the same movie and therefore the recommendation list must be adjusted as per the group interest. PolyLens (extension of MovieLens) demonstrates one of such recommender system. Similarly, in case of movie domain, we have seen MusicFX which recommends music as per the group interest. One basic question arises: Why not we can just average the recommendations generated by traditional recommendation approaches? The answer to this lies in a simple philosophy that: An individual decision might be influenced by other individuals. This phenomenon is referred as social influence. For example: Suppose a group of users are having a house warming party and some of the friends are not enjoying the song being played there. In such a scenario, averaging won't be an efficient to generalize the interest of the group. Therefore, while designing a recommendation system for group, it is utmost important to consider the social factors. In order to avoid noise and outliers while predicting the interest of group, median value can be used in place of mean. Different strategies that can be used to perform averaging in content-based RS or CF recommender system includes: *i.)* Weighted Averaging *ii.)* Average without misery, and *iii.)* Least misery strategy.

Multi-criteria RS

The users in any system might have shown interest in any particular item based on some criteria. Different user follows different criteria for showing their interest in any particular item. The recommender system should be able to capture the criteria on basis of which the user has shown his/her interest. This type of recommender system is inherent from Knowledge-based RS. Some of the methods that can be used to analyze the different criteria includes: Ensemble methods, neighborhood approach, etc. Table 8.2 defines an instance of multi- criteria systems. Here User A, B, C, D, E are 5 different users who showed their interest in any particular movie based on different criteria like: movie genre, visual effects, and movie cast. So, it is the primary job of recommender system to analyse these criteria before generating the recommendation for any individual.

Table 8.2 Multi-criteria systems

User_ID	Movie Cast	Effects	Genre
A	7	5	3
B	2	9	6
C	3	8	4
D	6	3	7
E	9	4	1

Active Learning

To predict which items in the consumer is most likely to be interested, an adaptive learning system acquires scores for only a select range of items. In a real-time scenario most of the rating dataset are highly sparse in nature, this gives rise to cold-start item and cold-start user problem. The recommender system needs to perform an active learning by gaining information about the items of which he is having no information. It is still one of the emerging areas of research with respect to collaborative filtering systems. Two widely used active learning-based strategies are: Performance-based and Heterogeneity-based models.

The ultimate aim of heterogeneity-based systems is to determine which user/item combination should be queried before executing the query. It's model- dependent to decide about whether uncertainty is high or low. The primary aim of querying for reviews is to improve estimation while lowering the uncertainty of future results. In models where performance is evaluated, the pieces of information are queried until precision on the residual entries gives the best estimated performance or the most likely certainty after the latest ranking is used. We must wait until all of the prediction are complete before calculating the rating. For comparison purposes, in uncertainty models, note that they rely on the state of the current requested instance, while performance models concern the effect that adding additional details would have on the accuracy of previously unknown entries. To compute the predicted score, you first have to decide the ranking in order to make the computation of this expected output possible via Bayesian methods.

Privacy

The gathering of input from different users is important for effective CF applications. In CF framework of recommender system, users are required to provide scores. User desires, political views, and so forth are revealed in these assessments. Where users have the option to rank things in public, they tend to be less considerate with their details. Some of the well-known techniques to preserve privacy includes privacy to be done at: collection time or publication time. The first model has better privacy safeguards because the individual's scores are not held in any specific manner. In some scenarios, the ranking is stored on an aggregated level. Thus, the solution offers better privacy protection. It is, on the other hand, more difficult to incorporate CF for such data collection approaches. This is either the data is influenced by high degree of variance, or the underlying method of collecting data has been altered to an aggregate form of considerable variation.

8.3 Applications of Deep Recommender Systems

This section highlights the diversified domain of recommender system application. Some of the well-known application of recommender systems are as follows:

1. News personalization
2. Advertising
3. Portal contents
4. Query recommendations
5. Reciprocal RS
6. Sequential modelling, etc.

To help the implementation of a wide range of product use cases such as music, film, social network, etc., recommender mechanisms have not only remained an active approach, but also have become difficult task. Newer experiments seek to classify hidden associations between knowledge using sophisticated computational models including deep networks and function learning. The key drawbacks of traditional approach to the production of recommendation-based features are their resource-intensive and time-consuming design and the reduction of recommendations in the space of learning; although the approaches using deep learning are powered by data, and can accurately derive key characteristics from raw data.

Looking ahead, we anticipate the following potential extensions of the recommender system:

1. Developing new public datasets that contain rich information about different domains, such as academic article metadata, or e-commerce content where the users do more than simply search or purchase and retrieve information.
2. Some of the implementation encourages the usage of users with different levels of experience to have useful feedback about how the recommender method works in different situations, which will lead to further under- standing of how it functions work in diverse use cases. Deep learning has been shown in just a handful of all the above initiatives to date. Although these additional changes need to be experienced in real-world conditions as well. It may be achieved through post-employment review of a company recommender systems through deep integration.
3. As far as deep learning models are concerned, there is a need for a statistical analysis that contrasts and measures all of deep learning techniques on the same standard datasets.
4. In comparison to conventional approaches, which depend on shallow layers of processing, which often work poorly, applying DL to RS increases their efficiency, but has a heavier computational load. Computational parameters often vary depending on the position of the consumer and item encounters significantly influence the model's overall complexity. Users' need to communicate with the data at the lower levels will be saved from the computing expense, allowing significant portions of data to be cached at the data layer. However, if the relationship exists at the lower levels, less computing may be used by several users, as usually happens, there is no re-usability of computational resources through users. Prerequisite for practical DL-based recommendation algorithms to be effectively applied in the real world, researchers must also fulfil the criteria for scaling the model to serve millions of users, and items.
5. Since deep learning algorithms are fundamentally hungry for input, they are more likely to match too tightly to their training sets. User input is much more likely to be scarce be-

cause people favour extended expression rather than abbreviated expression. By expanding on similar activities, this strategy and using an inductive bias, and embracing shared representations, researchers may train new experts to acquire and leverage different and extend their expertise simultaneously.

6. Transparency/Explainability is one of the important issues for recommender schemes and in general governments, legislative authorities, as well as commercial parties are eager to know why a particular solution was suggested at the moment. Models based on deep learning are considered a black box, with no understanding about how they make decisions. Users and companies will feel assured by the justification of the model's suggestion mechanism. In Chapter 4, we have tried to incorporate the explainability component with a deep belief network. But still lot more research needs to be done in this direction to further improvise the performance of model on all benchmark datasets.

8.4 Conclusion

In this chapter, we highlighted some of the recent advancements or recommender system along with possible extensions that can be done by using deep learning approaches. We also contrasted some of the key challenges of existing recommendation systems. The readers are advised to follow deep learning book written by Ian Goodfellow to get better insights of different deep learning approaches. With all this information, I would conclude this book with a beautiful quote said by SpaceX founder Elon Musk:

"I read books and talked to people. I mean that's kind of how one learns anything. There's lots of great books out there & lots of smart people"

— Elon Musk

Keep Reading!!
Keep Growing!!

Bibliography

[10] Recommender systems: an introduction. Cambridge University Press, 2010.

[AB20] Anand, R. and Beel, J. 'Auto-surprise: An automated recommender-system (autorecsys) library with tree of parzens estimator (tpe) optimization'. In: Fourteenth ACM Conference on Recommender Systems. 2020, pp. 585–587.

[ABC20] Ahmadianfar, I., Bozorg-Haddad, O. and Chu, X. 'Gradient- based optimizer: A new Metaheuristic optimization algorithm'. In: Information Sciences vol. 540 (2020), pp. 131–159.

[Agg+16] Aggarwal, C. C. et al. Recommender systems. Vol. 1. Springer, 2016.

[Ahm98] Ahmad Wasfi, A. M. 'Collecting user access patterns for building user profiles and collaborative filtering'. In: Proceedings of the 4th international conference on Intelligent user interfaces. 1998, pp. 57–64.

[Ahn08] Ahn, H. J. 'A new similarity measure for collaborative filtering to alleviate the new user cold-starting problem'. In: Information Sciences vol. 178, no. 1 (2008), pp. 37–51.

[Ala+17] Alashkar, T. et al. 'Examples-Rules Guided Deep Neural Network for Makeup Recommendation.' In: AAAI. 2017, pp. 941–947.

[AN20] Anand, P. B. and Nath, R. 'Content-Based Recommender Systems'. In: Recommender System with Machine Learning and Artificial Intelligence: Practical Tools and Applications in Medical, Agricultural and Other Industries (2020), p. 167.

[Ask16] Askarzadeh, A. 'A novel metaheuristic method for solving constrained engineering optimization problems: crow search algorithm'. In: Computers & Structures vol. 169 (2016), pp. 1–12.

[AT05] Adomavicius, G. and Tuzhilin, A. 'Toward the next generation of recommender systems: A survey of the state-of-the-art and possible extensions'. In: IEEE transactions on knowledge and data engineering vol. 17, no. 6 (2005), pp. 734–749.

[Bah+15] Bahrampour, S. et al. 'Comparative study of deep learning software frameworks'. In: arXiv preprint arXiv:1511.06435 (2015).

[BBM16] Bansal, T., Belanger, D. and McCallum, A. 'Ask the gru: Multi-task learning for deep text recommendations'. In: Proceedings of the 10th ACM Conference on Recommender Systems. 2016, pp. 107–114.

[BHB11] Bizer, C., Heath, T. and Berners-Lee, T. 'Linked data: The story so far'. In: Semantic services, interoperability and web applications: emerging concepts. IGI global, 2011, pp. 205–227.

[BHC+98] Basu, C., Hirsh, H., Cohen, W. et al. 'Recommendation as classification: Using social and content-based information in recommendation'. In: Aaai/iaai. 1998, pp. 714–720.

[BHY97] Burke, R. D., Hammond, K. J. and Yound, B. 'The FindMe approach to assisted browsing'. In: IEEE Expert vol. 12, no. 4 (1997), pp. 32–40.

[BKW17] Berg, R. v. d., Kipf, T. N. and Welling, M. 'Graph convolutional matrix completion'. In: arXiv preprint arXiv:1706.02263 (2017).

[Bob+12] Bobadilla, J. et al. 'Collaborative filtering based on significances. In: Information Sciences vol. 185, no. 1 (2012), pp. 1–17.

[Bob+13] Bobadilla, J. et al. 'Recommender systems survey'. In: Knowledge- based systems vol. 46 (2013), pp. 109–132.

[BOH12] Bobadilla, J., Ortega, F. and Hernando, A. 'A collaborative filtering similarity measure based on singularities. In: Information Processing & Management vol. 48, no. 2 (2012), pp. 204–217.

[BP99] Billsus, D. and Pazzani, M. J. 'A hybrid user model for news story classification'. In: Um99 user modeling. Springer, 1999, pp. 99–108.

[BSB10] Bobadilla, J., Serradilla, F. and Bernal, J. 'A new collaborative filtering metric that improves the behavior of recommender systems'. In: Knowledge-Based Systems vol. 23, no. 6 (2010), pp. 520–528.

[Bur02] Burke, R. 'Hybrid recommender systems: Survey and experiments'. In: User modeling and user-adapted interaction vol. 12, no. 4 (2002), pp. 331–370.

[Car+12] Carrer-Neto, W. et al. 'Social knowledge-based recommender system. Application to the movies domain'. In: Expert Systems with applications vol. 39, no. 12 (2012), pp. 10990–11000.

[CAS16] Covington, P., Adams, J. and Sargin, E. 'Deep neural networks for youtube recommendations'. In: *Proceedings of the 10th ACM conference on recommender systems*. 2016, pp. 191–198.

[CÇÖ17] Cengil, E., Çınar, A. and Özbay, E. 'Image classification with caffe deep learning framework'. In: *2017 International Conference on Computer Science and Engineering (UBMK)*. IEEE. 2017, pp. 440–444.

[CH05] Carreira-Perpinan, M. A. and Hinton, G. E. 'On contrastive divergence learning.' In: *Aistats*. Vol. 10. Citeseer. 2005, pp. 33–40.

[Che+15] Chen, T. et al. 'Mxnet: A flexible and efficient machine learning library for heterogeneous distributed systems'. In: *arXiv preprint arXiv:1512.01274* (2015).

[Che+17a] Chen, C. et al. 'Locally connected deep learning framework for industrial-scale recommender systems'. In: *Proceedings of the 26th International Conference on World Wide Web Companion*. 2017, pp. 769–770.

[Che+17b] Chen, X. et al. 'Personalized key frame recommendation'. In: Proceedings of the 40th International ACM SIGIR Conference on Research and Development in Information Retrieval. 2017, pp. 315–324.

[Che+20] Chen, J. et al. 'A novel recommender algorithm based on graph embedding and diffusion sampling'. In: Concurrency and Computation: Practice and Experience (2020), e5664.

[Chu+17] Chung, Y. et al. 'Comparison of deep learning frameworks: about theano, tensorflow, and cognitive toolkit'. In: Journal of Intelligence and Information Systems vol. 23, no. 2 (2017), pp. 1–17

[CJ18] Campos, P. N. and Jiménez, A. S. M. 'Attentive Collaborative Fil- tering: Multimedia Recommendation with Item-and Component- Level Attention'. In: (2018).

[CK18] Chae, D.-K. and Kim, S.-W. 'Adversarial Training of Deep Autoen- coders Towards Recommendation Tasks'. In: 2018 International Conference on Network Infrastructure and Digital Content (IC- NIDC). IEEE. 2018, pp. 91–95.

[CL20] Chou, C.-L. and Lu, T.-Y. 'A hybrid-feedback recommender sys- tem for employment websites'. In: Journal of Ambient Intelligence and Humanized Computing (2020), pp. 1–11.

[Dai+16] Dai, H. et al. 'Recurrent coevolutionary latent feature processes for continuous-time recommendation'. In: Proceedings of the 1st Workshop on Deep Learning for Recommender Systems. 2016, pp. 29–34.

[DCK20] Dara, S., Chowdary, C. R. and Kumar, C. 'A survey on group recommender systems'. In: Journal of Intelligent Information Systems vol. 54, no. 2 (2020), pp. 271–295.

[Den15] Deng, F. 'Utility-based Recommender Systems Using Implicit Utility and Genetic Algorithm'. In: 2015 International Conference on Mechatronics, Electronic, Industrial and Control Engineering (MEIC-15). Atlantis Press. 2015.

[DGM08] Debnath, S., Ganguly, N. and Mitra, P. 'Feature weighting in content based recommendation system using social network analysis'. In: Proceedings of the 17th international conference on World Wide Web. 2008, pp. 1041–1042.

[DK17] Dhiman, G. and Kumar, V. 'Spotted hyena optimizer: a novel bio-inspired based metaheuristic technique for engineering applications'. In: Advances in Engineering Software vol. 114 (2017), pp. 48–70.

[DR15] Dziugaite, G. K. and Roy, D. M. 'Neural network matrix factorization'. In: arXiv preprint arXiv:1511.06443 (2015).

[EF17] Ebesu, T. and Fang, Y. 'Neural citation network for context- aware citation recommendation'. In: Proceedings of the 40th international ACM SIGIR conference on research and development in information retrieval. 2017, pp. 1093–1096.

[EK19] Eftimov, T. and Korošec, P. 'A novel statistical approach for comparing meta-heuristic stochastic optimization algorithms according to the distribution of solutions in the search space'. In: Information Sciences vol. 489 (2019), pp. 255–273.

[EKS17] Eftimov, T., Korošec, P. and Seljak, B. K. 'A novel approach to statistical comparison of meta-heuristic stochastic optimization algorithms using deep statistics'. In: Information Sciences vol. 417 (2017), pp. 186–215.

[ESH15] Elkahky, A. M., Song, Y. and He, X. 'A multi-view deep learning approach for cross domain user modeling in recommendation systems'. In: Proceedings of the 24th international conference on world wide web. 2015, pp. 278–288.

[Fan+19a] Fan, W. et al. 'Deep adversarial social recommendation'. In: arXiv preprint arXiv:1905.13160 (2019).

[Fan+19b] Fan, W. et al. 'Deep social collaborative filtering'. In: Proceedings of the 13th ACM Conference on Recommender Systems. 2019, pp. 305–313.

[Fan+19c] Fan, W. et al. 'Graph neural networks for social recommendation'. In: The World Wide Web Conference. 2019, pp. 417–426.

[Far+20] Faramarzi, A. et al. 'Marine Predators Algorithm: A nature- inspired metaheuristic'. In: Expert Systems with Applications vol. 152 (2020), p. 113377.

[FB11] Farzan, R. and Brusilovsky, P. 'Encouraging user participation in a course recommender system: An impact on user behavior'. In: Computers in Human Behavior vol. 27, no. 1 (2011), pp. 276–284.

[Gar+09] Garcıa, S. et al. 'A study on the use of non-parametric tests for analyzing the evolutionary algorithms' behaviour: a case study on the CEC'2005 special session on real parameter optimization'. In: Journal of Heuristics vol. 15, no. 6 (2009), p. 617.

[Ger+10] Gershman, A. et al. 'A decision tree based recommender system'. In: 10th International Conferenceon Innovative Internet Community Systems (I2CS)–Jubilee Edition 2010– (2010).

[GK20] Gupta, S. and Kant, V. 'Credibility score based multi-criteria recommender system'. In: Knowledge-Based Systems (2020), p. 105756.

[GLD18] Gulzar, Z., Leema, A. A. and Deepak, G. 'Pcrs: Personalized course recommender system based on hybrid approach'. In: Procedia Computer Science vol. 125 (2018), pp. 518–524.

[GN13] Georgiev, K. and Nakov, P. 'A non-iid framework for collaborative filtering with restricted boltzmann machines'. In: International conference on machine learning. 2013, pp. 1148–1156.

[Goo+16] Goodfellow, I. et al. Deep learning. Vol. 1. 2. MIT press Cambridge, 2016.

[GP17] Gulli, A. and Pal, S. Deep learning with Keras. Packt Publishing Ltd, 2017.

[GR12] Gantz, J. and Reinsel, D. 'The digital universe in 2020: Big data, bigger digital shadows, and biggest growth in the far east'. In: IDC iView: IDC Analyze the future vol. 2007, no. 2012 (2012), pp. 1–16.

[GSM14] Gogna, A., Shukla, A. and Majumdar, A. 'Matrix recovery using split bregman'. In: 2014 22nd International Conference on Pattern Recognition. IEEE. 2014, pp. 1031–1036.

[GZ16] Gong, Y. and Zhang, Q. 'Hashtag recommendation using attention- based convolutional neural network.' In: IJCAI. 2016, pp. 2782– 2788.

[GZY13] Guo, G., Zhang, J. and Yorke-Smith, N. 'A Novel Bayesian Similarity Measure for Recommender Systems.' In: IJCAI. Vol. 13. 2013, pp. 2619–2625.

[He+17] He, X. et al. 'Neural collaborative filtering'. In: Proceedings of the 26th international conference on world wide web. 2017, pp. 173– 182.

[He+18a] He, X. et al. 'Adversarial personalized ranking for recommendation'. In: The 41st International ACM SIGIR Conference on Research & Development in Information Retrieval. 2018, pp. 355– 364.

[He+18b] He, X. et al. 'Outer product-based neural collaborative filtering'. In: arXiv preprint arXiv:1808.03912 (2018).

[Hei+19] Heidari, A. A. et al. 'An efficient hybrid multilayer perceptron neural network with grasshopper optimization'. In: Soft Computing vol. 23, no. 17 (2019), pp. 7941–7958.

[Her+17] Herlocker, J. L. et al. 'An algorithmic framework for performing collaborative filtering'. In: ACM SIGIR Forum. Vol. 51. 2. ACM New York, NY, USA. 2017, pp. 227–234.

[Hid+15] Hidasi, B. et al. 'Session-based recommendations with recurrent neural networks'. In: arXiv preprint arXiv:1511.06939 (2015).

[Hin02] Hinton, G. E. 'Training products of experts by minimizing contrastive divergence'. In: Neural computation vol. 14, no. 8 (2002), pp. 1771–1800.

[Hin12] Hinton, G. E. 'A practical guide to training restricted Boltzmann machines'. In: Neural networks: Tricks of the trade. Springer, 2012, pp. 599–619.

[Hor91] Hornik, K. 'Approximation capabilities of multilayer feedforward networks'. In: Neural networks vol. 4, no. 2 (1991), pp. 251–257.

[Hsi+17] Hsieh, C.-K. et al. 'Collaborative metric learning'. In: Proceedings of the 26th international conference on world wide web. 2017, pp. 193–201.

[Hu+15] Hu, Y.-C. et al. 'A fuzzy similarity measure for collaborative filtering using nonadditive grey relational analysis.' In: Journal of Grey System vol. 27, no. 2 (2015).

[HZN15] Hosseini-Asl, E., Zurada, J. M. and Nasraoui, O. 'Deep learning of part-based representation of data using sparse autoencoders with nonnegativity constraints'. In: IEEE transactions on neural networks and learning systems vol. 27, no. 12 (2015), pp. 2486–2498.

[IBH15] Ilany, A., Booms, A. S. and Holekamp, K. E. 'Topological effects of network structure on long-term social network dynamics in a wild mammal'. In: Ecology letters vol. 18, no. 7 (2015), pp. 687–695.

[IS20] Ibrahim, M. S. and Saidu, C. I. 'Recommender Systems: Al- gorithms, Evaluation and Limitations'. In: Journal of Advances in Mathematics and Computer Science (2020), pp. 121–137.

[JCB18] Jakomin, M., Curk, T. and Bosnić, Z. 'Generating interdependent data streams for recommender systems'. In: Simulation Modelling Practice and Theory vol. 88 (2018), pp. 1–16.

[JD19] Jain, P. and Dixit, V. S. 'Recommendations with context aware framework using particle swarm optimization and unsupervised learning'. In: Journal of Intelligent & Fuzzy Systems vol. 36, no. 5 (2019), pp. 4479–4490.

[JDF19] Jiang, S., Ding, Z. and Fu, Y. 'Heterogeneous recommendation via deep low-rank sparse collective factorization'. In: IEEE transactions on pattern analysis and machine intelligence vol. 42, no. 5 (2019), pp. 1097–1111.

[JE09] Jamali, M. and Ester, M. 'Trustwalker: a random walk model for combining trust-based and item-based recommendation'. In: Proceedings of the 15th ACM SIGKDD international conference on Knowledge discovery and data mining. 2009, pp. 397–406.

[JE10] Jamali, M. and Ester, M. 'A matrix factorization technique with trust propagation for recommendation in social networks'. In: Proceedings of the fourth ACM conference on Recommender systems. 2010, pp. 135–142.

[JSS19] Javidi, A., Salajegheh, E. and Salajegheh, J. 'Enhanced crow search algorithm for optimum design of structures'. In: Applied Soft Computing vol. 77 (2019), pp. 274–289.

[KBV09] Koren, Y., Bell, R. and Volinsky, C. 'Matrix factorization techniques for recommender systems'. In: Computer vol. 42, no. 8 (2009), pp. 30–37.

[KKB20] Kiran, R., Kumar, P. and Bhasker, B. 'DNNRec: A novel deep learning based hybrid recommender system'. In: Expert Systems with Applications vol. 144 (2020), p. 113054.

[KM18] Kant, S. and Mahara, T. 'Merging user and item based collabor- ative filtering to alleviate data sparsity'. In: International Journal of System Assurance Engineering and Management vol. 9, no. 1 (2018), pp. 173–179.

[Kor08] Koren, Y. 'Factorization meets the neighborhood: a multifaceted collaborative filtering model'. In: Proceedings of the 14th ACM SIGKDD international conference on Knowledge discovery and data mining. 2008, pp. 426–434.

[Kum+17] Kumar, V. et al. 'Proximal maximum margin matrix factorization for collaborative filtering'. In: Pattern Recognition Letters vol. 86 (2017), pp. 62–67.

[KV17] Katarya, R. and Verma, O. P. 'An effective collaborative movie recommender system with cuckoo search'. In: Egyptian Informatics Journal vol. 18, no. 2 (2017), pp. 105–112.

[Lan+19] Lang, S. et al. 'Wekadeeplearning4j: A deep learning package for weka based on deeplearning4j'. In: Knowledge-Based Systems vol. 178 (2019), pp. 48–50.

[Lee+13] Lee, J. et al. 'Local low-rank matrix approximation'. In: Inter- national conference on machine learning. PMLR. 2013, pp. 82- 90.

[Lee+16] Lee, H. et al. 'Quote recommendation in dialogue using deep neural network'. In: Proceedings of the 39th International ACM SIGIR conference on Research and Development in Information Retrieval. 2016, pp. 957–960.

[Lee+18] Lee, J. et al. 'Collaborative deep metric learning for video understanding'. In: Proceedings of the 24th ACM SIGKDD International Conference on Knowledge Discovery & Data Mining. 2018, pp. 481–490.

[Lei+16] Lei, C. et al. 'Comparative deep learning of hybrid representations for image recommendations'. In: Proceedings of the IEEE Conference on Computer Vision and Pattern Recognition. 2016, pp. 2545–2553.

[Li+16] Li, Y. et al. 'Hashtag recommendation with topical attention-based LSTM'. In: Coling. 2016.

[Li+17] Li, P. et al. 'Neural rating regression with abstractive tips generation for recommendation'. In: Proceedings of the 40th International ACM SIGIR conference on Research and Development in Information Retrieval. 2017, pp. 345–354.

[Lia+18] Liang, D. et al. 'Variational autoencoders for collaborative filtering'. In: Proceedings of the 2018 World Wide Web Conference. 2018, pp. 689–698.

[Liu+15a] Liu, J. et al. 'Domain-sensitive recommendation with user-item subgroup analysis'. In: IEEE Transactions on Knowledge and Data Engineering vol. 28, no. 4 (2015), pp. 939–950.

[Liu+15b] Liu, X. et al. 'Item category aware conditional restricted boltzmann machine based recommendation'. In: International Conference on Neural Information Processing. Springer. 2015, pp. 609–616.

[Liu+20] Liu, H. et al. 'Automated Embedding Size Search in Deep Recommender Systems'. In: Proceedings of the 43rd International ACM SIGIR Conference on Research and Development in Information Retrieval. 2020, pp. 2307–2316.

[Liu18] Liu, J. Y. 'A Survey of Deep Learning Approaches for Recommendation Systems'. In: Journal of Physics: Conference Series. Vol. 1087. 6. 2018.

[LKF15] Li, S., Kawale, J. and Fu, Y. 'Deep collaborative filtering via marginalized denoising auto-encoder'. In: Proceedings of the 24th ACM International on Conference on Information and Knowledge Management. 2015, pp. 811–820.

[LKH14] Lika, B., Kolomvatsos, K. and Hadjiefthymiades, S. 'Facing the cold start problem in recommender systems'. In: *Expert Systems with Applications* vol. 41, no. 4 (2014), pp. 2065–2073.

[LS17] Li, X. and She, J. 'Collaborative variational autoencoder for recommender systems'. In: *Proceedings of the 23rd ACM SIGKDD international conference on knowledge discovery and data mining*.2017, pp. 305–314.

[LSY03] Linden, G., Smith, B. and York, J. 'Amazon. com recommend- ations: Item-to-item collaborative filtering'. In: *IEEE Internet computing* vol. 7, no. 1 (2003), pp. 76–80.

[Ma+11] Ma, H. et al. 'Recommender systems with social regularization'. In: *Proceedings of the fourth ACM international conference on Web search and data mining*. 2011, pp. 287–296.

[Ma+19] Ma, J. et al. 'Learning disentangled representations for recommend- ation'. In: *Advances in Neural Information Processing Systems*.2019, pp. 5711–5722.

[Mir+99] Miranda, T. et al. 'Combining content-based and collaborative filters in an online newspaper'. In: *In Proceedings of ACM SIGIR Workshop on Recommender Systems*. Citeseer. 1999.

[MJ16] Menhaj, M. B. and Jamalzehi, S. 'Scalable user similarity es- timation based on fuzzy proximity for enhancing accuracy of collaborative filtering recommendation'. In: *2016 4th Interna- tional Conference on Control, Instrumentation, and Automation (ICCIA)*. IEEE. 2016, pp. 220–225.

[MKI19] Mohamed, M. H., Khafagy, M. H. and Ibrahim, M. H. 'Recom- mender systems challenges and solutions survey'. In: *2019 Interna- tional Conference on Innovative Trends in Computer Engineering (ITCE)*. IEEE. 2019, pp. 149–155.

[MM20] Mandal, S. and Maiti, A. 'Explicit feedback meet with implicit feedback in GPMF: a generalized probabilistic matrix factorization model for recommendation'. In: *Applied Intelligence* (2020), pp. 1–24.

[Moh+19] Mohammadpour, T. et al. 'Efficient clustering in collaborative filtering recommender system: Hybrid method based on genetic algorithm and gravitational emulation local search algorithm'. In: *Genomics* vol. 111, no. 6 (2019), pp. 1902–1912.

[MR00] Mooney, R. J. and Roy, L. 'Content-based book recommending using learning for text categorization'. In: *Proceedings of the fifth ACM conference on Digital libraries*. 2000, pp. 195–204.

[MS07] Mnih, A. and Salakhutdinov, R. R. 'Probabilistic matrix factor- ization'. In: *Advances in neural information processing systems* vol. 20 (2007), pp. 1257–1264.

[MT17] Majumdar, A. and Tripathi, A. 'Asymmetric stacked autoencoder'. In: *2017 International Joint Conference on Neural Networks (IJCNN)*. IEEE. 2017, pp. 911–918.

[MTF20] Milano, S., Taddeo, M. and Floridi, L. 'Recommender systems and their ethical challenges'. In: *AI & SOCIETY* (2020), pp. 1–11.

[Muh+20] Muhairat, M. et al. 'An Intelligent Recommender System Based on Association Rule Analysis for Requirement Engineering.' In: *J. UCS* vol. 26, no. 1 (2020), pp. 33–49.

[Mus+18] Musto, C. et al. 'Deep content-based recommender systems exploiting recurrent neural networks and linked open data'. In: *Adjunct Publication of the 26th Conference on User Modeling, Adaptation and Personalization*. 2018, pp. 239–244.

[MW15] Marz, N. and Warren, J. *Big Data: Principles and best practices of scalable real-time data systems*. New York; Manning Publications Co., 2015.

[MZ20] Mu, R. and Zeng, X. 'Auxiliary Stacked Denoising Autoencoder based Collaborative Filtering Recommendation.' In: *KSII Trans- actions on Internet & Information Systems* vol. 14, no. 6 (2020).

[Nat+20] Natarajan, S. et al. 'Resolving data sparsity and cold start problem in collaborative filtering recommender system using linked open data'. In: *Expert Systems with Applications* vol. 149 (2020), p. 113248.

[NBF19] Nikzad-Khasmakhi, N., Balafar, M. and Feizi-Derakhshi, M. R. 'The state-of-the-art in expert recommendation systems'. In: *Engineering Applications of Artificial Intelligence* vol. 82 (2019), pp. 126–147.

[Ngu+14] Nguyen, T. T. et al. 'Exploring the filter bubble: the effect of using recommender systems on content diversity'. In: *Proceedings of the 23rd international conference on World wide web*. 2014, pp. 677–686.

[Oku+17] Okura, S. et al. 'Embedding-based news recommendation for millions of users'. In: *Proceedings of the 23rd ACM SIGKDD International Conference on Knowledge Discovery and Data Mining*. 2017, pp. 1933–1942.

[OP13] Ozsoy, M. G. and Polat, F. 'Trust based recommendation systems'. In: *Proceedings of the 2013 IEEE/ACM International Conference on Advances in Social Networks Analysis and Mining*. 2013,pp. 1267–1274.

[Par+12] Park, D. H. et al. 'A literature review and classification of recom- mender systems research'. In: *Expert systems with applications* vol. 39, no. 11 (2012), pp. 10059–10072.

[PHY20] Pan, Y., He, F. and Yu, H. 'A correlative denoising autoencoder to model social influence for top-N recommender system'. In: *Frontiers of Computer Science* vol. 14, no. 3 (2020), p. 143301.

[Qua+20] Quan, Y. et al. 'Collaborative deep learning for super-resolving blurry text images'. In: *IEEE Transactions on Computational Imaging* (2020).

[RK16] Rawat, Y. S. and Kankanhalli, M. S. 'ConTagNet: Exploiting user context for image tag recommendation'. In: *Proceedings of the 24th ACM international conference on Multimedia*. 2016, pp. 1102–1106.

[Rob+14] Robillard, M. et al. *Recommendation Systems in Software Engin- eering*. Springer-Verlag Berlin Heidelberg, 2014.

[RRS11] Ricci, F., Rokach, L. and Shapira, B. 'Introduction to recom- mender systems handbook'. In: *Recommender systems handbook*. Springer, 2011, pp. 1–35.

[SC00] Smyth, B. and Cotter, P. 'A personalised TV listings service for the digital TV age'. In: *Knowledge-Based Systems* vol. 13, no. 2-3 (2000), pp. 53–59.

[SD19a] Sinha, B. B. and Dhanalakshmi, R. 'Evolution of recommender paradigm optimization over time'. In: *Journal of King Saud University-Computer and Information Sciences* (2019).

[SD19b] Sinha, B. B. and Dhanalakshmi, R. 'Evolution of recommender system over the time'. In: *Soft Computing* vol. 23, no. 23 (2019),pp. 12169–12188.

[SD20] Sinha, B. B. and Dhanalakshmi, R. 'A recommender system based on a new similarity metric and upgraded crow search algorithm'. In: *Journal of Intelligent & Fuzzy Systems*, no. Preprint (2020),pp. 1–16.

[Sed+15] Sedhain, S. et al. 'Autorec: Autoencoders meet collaborative filtering'. In: *Proceedings of the 24th international conference on World Wide Web*. 2015, pp. 111–112.

[She+16] Shen, X. et al. 'Automatic recommendation technology for learning resources with convolutional neural network'. In: *2016 International Symposium on Educational Technology (ISET)*. IEEE. 2016, pp. 30–34.

[SKK01] Schwab, I., Kobsa, A. and Koychev, I. 'Learning user interests through positive examples using content analysis and collaborative filtering'. In: *Internal Memo, GMD, St. Augustin, Germany* (2001).

[SKR01] Schafer, J. B., Konstan, J. A. and Riedl, J. 'E-commerce recommendation applications'. In: *Data mining and knowledge discovery* vol. 5, no. 1-2 (2001), pp. 115–153.

[SMH07] Salakhutdinov, R., Mnih, A. and Hinton, G. 'Restricted Boltzmann machines for collaborative filtering'. In: *Proceedings of the 24th international conference on Machine learning*. 2007, pp. 791–798.

[Sun+15] Sun, Z. et al. 'Recommender systems based on social networks'. In: *Journal of Systems and Software* vol. 99 (2015), pp. 109–119.

[TA19] Thies, J. and Alimohammad, A. 'Compact and low-power neural spike compression using undercomplete autoencoders'. In: *IEEE Transactions on Neural Systems and Rehabilitation Engineering* vol. 27, no. 8 (2019), pp. 1529–1538.

[TLH18] Tay, Y., Luu, A. T. and Hui, S. C. 'Multi-pointer co-attention networks for recommendation'. In: *Proceed-*

ings of the 24th ACM SIGKDD International Conference on Knowledge Discovery & Data Mining. 2018, pp. 2309–2318.

[TRD20] Tarnowska, K., Ras, Z. W. and Daniel, L. *Recommender System for Improving Customer Loyalty*. Springer, 2020.

[TW18] Tang, J. and Wang, K. 'Personalized top-n sequential recommend- ation via convolutional sequence embedding'. In: *Proceedings of the Eleventh ACM International Conference on Web Search and Data Mining*. 2018, pp. 565–573.

[VDS13] Van den Oord, A., Dieleman, S. and Schrauwen, B. 'Deep content- based music recommendation'. In: *Advances in neural information processing systems*. 2013, pp. 2643–2651.

[Vie18] Vie, J.-J. 'Deep factorization machines for knowledge tracing'. In: arXiv preprint arXiv:1805.00356 (2018).

[Vin+08] Vincent, P. et al. 'Extracting and composing robust features with denoising autoencoders'. In: *Proceedings of the 25th international conference on Machine learning*. 2008, pp. 1096–1103.

[Wan+17a] Wang, J. et al. 'Irgan: A minimax game for unifying generative and discriminative information retrieval models'. In: *Proceedings of the 40th International ACM SIGIR conference on Research and Development in Information Retrieval*. 2017, pp. 515–524.

[Wan+17b] Wang, S. et al. 'What your images reveal: Exploiting visual contents for point-of-interest recommendation'. In: *Proceedings of the 26th International Conference on World Wide Web*. 2017,pp. 391–400.

[Wan+18a] Wang, H. et al. 'Ripplenet: Propagating user preferences on the knowledge graph for recommender systems'. In: *Proceedings of the 27th ACM International Conference on Information and Knowledge Management*. 2018, pp. 417–426.

[Wan+18b] Wang, L. et al. 'Supervised reinforcement learning with recurrent neural network for dynamic treatment recommendation'. In: *Proceedings of the 24th ACM SIGKDD International Conference on Knowledge Discovery & Data Mining*. 2018, pp. 2447–2456.

[Wan+19] Wang, Y. et al. 'Assessing optimizer impact on dnn model sensitivity to adversarial examples'. In: *IEEE Access* vol. 7 (2019),pp. 152766–152776.

[Wan+20a] Wang, T.-H. et al. 'AutoRec: An Automated Recommender System'. In: *arXiv preprint arXiv:2007.07224* (2020).

[Wan+20b] Wang, W. et al. 'Denoising Implicit Feedback for Recommenda- tion'. In: *arXiv preprint arXiv:2006.04153* (2020).

[Wei+16] Wei, J. et al. 'Collaborative filtering and deep learning based hy- brid recommendation for cold start problem'. In: *2016 IEEE 14th Intl Conf on Dependable, Autonomic and Secure Computing, 14th Intl Conf on Pervasive Intelligence and Computing, 2nd Intl Conf on Big Data Intelligence and Computing and Cyber Science and Technology Congress (DASC/PiCom/DataCom/CyberSciTech)*. IEEE. 2016, pp. 874–877.

[Wig+19] Wigren, A. et al. 'Parameter elimination in particle Gibbs sampling'. In: *arXiv preprint arXiv:1910.14145* (2019).

[WL14] Wang, H. and Li, W.-J. 'Relational collaborative topic regression for recommender systems'. In: *IEEE Transactions on Knowledge and Data Engineering* vol. 27, no. 5 (2014), pp. 1343–1355.

[WSY15] Wang, H., Shi, X. and Yeung, D.-Y. 'Relational stacked denoising autoencoder for tag recommendation'. In: *Twenty-ninth AAAI conference on artificial intelligence*. Citeseer. 2015.

[Wu+16a] Wu, H. et al. 'Collaborative topic regression with social trust ensemble for recommendation in social media systems'. In: *Knowledge-Based Systems* vol. 97 (2016), pp. 111–122.

[Wu+16b] Wu, S. et al. 'Personal recommendation using deep recurrent neural networks in NetEase'. In: *2016 IEEE 32nd international conference on data engineering (ICDE)*. IEEE. 2016, pp. 1218– 1229.

[Wu+16c] Wu, Y. et al. 'Collaborative denoising auto-encoders for top- n recommender systems'. In: *Proceedings of the ninth ACM international conference on web search and data mining*. 2016, pp. 153–162.

[Wu+17] Wu, C.-Y. et al. 'Recurrent recommender networks'. In: *Proceed- ings of the tenth ACM international conference on web search and data mining*. 2017, pp. 495–503.

[WWY15] Wang, H., Wang, N. and Yeung, D.-Y. 'Collaborative deep learningfor recommender systems'. In: *Proceedings of the 21th ACM SIGKDD international conference on knowledge discovery and data mining*. 2015, pp. 1235–1244.

[WXY16] Wang, H., Xingjian, S. and Yeung, D.-Y. 'Collaborative recurrent autoencoder: Recommend while learning to fill in the blanks'. In: *Advances in Neural Information Processing Systems*. 2016, pp. 415–423.

[Yan+20] Yang, Y. et al. 'A group preference-based item similarity model: comparison of clustering techniques in ambient and context-aware recommender systems'. In: *Journal of Ambient Intelligence and Humanized Computing* vol. 11, no. 4 (2020), pp. 1441–1449.

[Yin+16] Ying, H. et al. 'Collaborative deep ranking: A hybrid pair-wise recommendation algorithm with implicit feedback'. In: *Pacific- asia conference on knowledge discovery and data mining*. Springer.2016, pp. 555–567.

[Yin+18] Ying, R. et al. 'Graph convolutional neural networks for web-scale recommender systems'. In: *Proceedings of the 24th ACM SIGKDD International Conference on Knowledge Discovery & Data Mining*. 2018, pp. 974–983.

[YL18] Yu, W. and Li, S. 'Recommender systems based on multiple social networks correlation'. In: *Future Generation Computer Systems*vol. 87 (2018), pp. 312–327.

[YM17] Yera, R. and Martinez, L. 'Fuzzy tools in recommender systems: A survey'. In: *International Journal of Computational Intelligence Systems* vol. 10, no. 1 (2017), pp. 776–803.

[Zha+16] Zhang, F. et al. 'Collaborative knowledge base embedding for recommender systems'. In: *Proceedings of the 22nd ACM SIGKDD international conference on knowledge discovery and data mining*. 2016, pp. 353–362.

[Zha+17] Zhang, Q. et al. 'Hashtag Recommendation for Multi-modal Mi- croblog Using Co-Attention Network.' In: *IJCAI*. 2017, pp. 3420– 3426.

[Zha+18a] Zhao, X. et al. 'Deep reinforcement learning for page-wise recommendations'. In: *Proceedings of the 12th ACM Conference on Recommender Systems*. 2018, pp. 95–103.

[Zha+18b] Zhao, X. et al. 'Recommendations with negative feedback via pairwise deep reinforcement learning'. In: *Proceedings of the 24th ACM SIGKDD International Conference on Knowledge Discovery & Data Mining*. 2018, pp. 1040–1048.

[Zha18] Zhang, Z. 'Improved adam optimizer for deep neural networks'. In: *2018 IEEE/ACM 26th International Symposium on Quality of Service (IWQoS)*. IEEE. 2018, pp. 1–2.

[Zhe+16] Zheng, Y. et al. 'A neural autoregressive approach to collaborative filtering'. In: *International Conference on Machine Learning*. PMLR. 2016, pp. 764–773.

[Zhe+18] Zheng, G. et al. 'DRN: A deep reinforcement learning framework for news recommendation'. In: *Proceedings of the 2018 World Wide Web Conference*. 2018, pp. 167–176.

[ZLJ20] Zhang, G., Liu, Y. and Jin, X. 'A survey of autoencoder-based recommender systems'. In: *Frontiers of Computer Science* (2020), pp. 1–21.

[ZNY17] Zheng, L., Noroozi, V. and Yu, P. S. 'Joint deep modeling of users and items using reviews for recommendation'. In: *Proceedings of the Tenth ACM International Conference on Web Search and Data Mining*. 2017, pp. 425–434.

[Zou+19] Zou, F. et al. 'A sufficient condition for convergences of adam and rmsprop'. In: *Proceedings of the IEEE/CVF Conference on Computer Vision and Pattern Recognition*. 2019, pp. 11127–11135.

[ZQX16] Zhao, G., Qian, X. and Xie, X. 'User-service rating prediction by exploring social users' rating behaviors'. In: *IEEE transactions on multimedia* vol. 18, no. 3 (2016), pp. 496–506.

[ZYX17] Zhang, S., Yao, L. and Xu, X. 'AutoSVD++ An Efficient Hybrid Collaborative Filtering Model via Contractive

Auto-encoders'. In: *Proceedings of the 40th International ACM SIGIR conference on Research and Development in Information Retrieval*. 2017, pp. 957–960.

www.ingramcontent.com/pod-product-compliance
Lightning Source LLC
LaVergne TN
LVHW052323200926
848481LV00002B/650
9781922617545